MICROBIAL GROWTH AND SURVIVAL IN EXTREMES OF ENVIRONMENT

THE SOCIETY FOR APPLIED BACTERIOLOGY
TECHNICAL SERIES NO. 15

MICROBIAL GROWTH AND SURVIVAL IN EXTREMES OF ENVIRONMENT

Edited by

G. W. GOULD

Unilever Research Laboratory, Colworth House,
Sharnbrook, Bedford MK44 1LQ

AND

JANET E. L. CORRY

The Metropolitan Police Forensic Science Laboratory,
109 Lambeth Road, London SE1 7LP

1980

ACADEMIC PRESS

A Subsidiary of Harcourt Brace Jovanovich, Publishers

LONDON · NEW YORK · TORONTO · SYDNEY · SAN FRANCISCO

ACADEMIC PRESS INC. (LONDON) LTD
24/28 OVAL ROAD
LONDON NW1

U.S. Edition Published by
ACADEMIC PRESS INC.
111 FIFTH AVENUE
NEW YORK, NEW YORK 10003

British Library Cataloguing in Publication Data

Microbial growth and survival in extremes of
   environment.—(Society for Applied Bacteriology.
   Technical series; 15 ISSN 0300–9610).
   1. Micro-organisms—Physiology
   2. Microbial ecology
   3. Adaptation (Physiology)
   I. Gould, Grahame Warwick   II. Corry, J. E. L.
   III. Series
   576′.15   QR97.A1   79–41561

   ISBN 0–12–293680–9

Printed in Great Britain by
Latimer Trend & Company Ltd, Plymouth

Contributors

S. R. ALCOCK, *Department of Bacteriology, The Medical School, Foresterhill, Aberdeen, UK*

G. D. ANAGNOSTOPOULOS, *Department of Microbiology, Queen Elizabeth College, University of London, Campden Hill, London W8 7AH, UK*

CELIA A. AYRES, *Campden Food Preservation R.A., Chipping Campden, Gloucestershire GL55 6LD, UK*

D. J. W. BARBER, *Department of Pharmacy, University of Manchester, Manchester M13 9PL, UK*

CATHERINE E. BAYLISS, *Agricultural Research Council, Food Research Institute, Colney Lane, Norwich NR4 7UA, UK*

P. G. BEAN, *Metal Box Ltd, Denchworth Road, Wantage OX12 9BP, UK*

CORALE L. BRIERLEY, *New Mexico Institute of Mining and Technology, Socorro, New Mexico 87801, USA*

J. A. BRIERLEY, *New Mexico Institute of Mining and Technology, Socorro, New Mexico 87801, USA*

K. L. BROWN, *Campden Food Preservation R.A., Chipping Campden, Gloucestershire GL55 6LD, UK*

M. H. BROWN, *Unilever Research Laboratory, Colworth House, Sharnbrook, Bedford MK44 1LQ, UK*

M. R. W. BROWN, *Department of Pharmacy, University of Aston in Birmingham, Gosta Green, Birmingham B4 7ET, UK*

SUSAN A. BUCKERIDGE, *Department of Applied Biochemistry and Nutrition, University of Nottingham School of Agriculture, Sutton Bonnington, Loughborough, Leicestershire LE12 5RD, UK*

H. DALLYN, *Metal Box Ltd, Denchworth Road, Wantage OX12 9BP, UK*

F. L. DAVIES, *National Institute for Research in Dairying, Shinfield, Reading RG2 9AT, UK*

G. DHAVISES, *Biology Department, Kasetsart University, Bangkok 9, Thailand*

C. S. DOW, *Department of Biological Sciences, University of Warwick, Coventry CV4 7AL, UK*

W. C. FALLOON, *Metal Box Ltd, Denchworth Road, Wantage OX12 9BP, UK*

A. Fox, *Department of Applied Biology and Food Science, Polytechnic of the South Bank, London SE1 OAA, UK*

R. J. Gilbert, *Food Hygiene Laboratory, Central Public Health Laboratory, Colindale Avenue, London NW9 5HT, UK*

W. D. Grant, *Department of Microbiology, The Medical School, University of Leicester, Leicester LE1 7RN, UK*

T. R. G. Gray, *Department of Biology, University of Essex, Wivenhoe Park, Colchester CO4 3SQ, UK*

P. Ineson, *Department of Biological Sciences, University of Exeter, Exeter EX4 4QG, UK*

B. Jarvis, *Leatherhead Food R.A., Randalls Road, Leatherhead, Surrey KT22 7RY, UK*

D. P. Kelly, *Department of Environmental Sciences, University of Warwick, Coventry CV4 7AL, UK*

Margaret Kendall, *Food Hygiene Laboratory, Central Public Health Laboratory, Colindale Avenue, London NW9 5HT, UK*

M. A. Kenward, *Department of Biological Sciences, Wolverhampton Polytechnic, Wolverhampton, West Midlands, UK*

R. G. Kroll, *Department of Microbiology, University of Aberdeen, Marischal College, Aberdeen AB9 1AS, UK*

J. Lacey, *Rothamsted Experimental Station, Harpenden, Hertfordshire AL5 2JQ, UK*

A. Lawrence, *Department of Biological Sciences, University of Warwick, Coventry CV4 7AL, UK*

H. L. M. Lelieveld, *Unilever Research Laboratory, Olivier van Nortlaan 120, Vlaardingen, Postbus 114, The Netherlands*

F. J. Ley, *Irradiated Products Ltd, Elgin Industrial Estate, Swindon, Wiltshire, UK*

D. W. Lovelock, *H. J. Heinz Co. Ltd, Hayes Park, Hayes, Middlesex UB4 8AL, UK*

T. Mayes, *Unilever Research Laboratory, Colworth House, Sharnbrook, Bedford MK44 1LQ, UK*

P. Neaves, *Leatherhead Food R.A., Randalls Road, Leatherhead, Surrey KT22 7RY, UK*

P. R. Norris, *Department of Environmental Sciences, University of Warwick, Coventry CV4 7AL, UK*

A. G. Perkin, *National Institute for Research in Dairying, Shinfield, Reading RG2 9AT, UK*

A. Seaman, *Department of Applied Biochemistry and Nutrition, University of Nottingham School of Agriculture, Sutton Bonnington, Loughborough, Leicestershire LE12 5RD, UK*

M. Shahamat, *Department of Applied Biochemistry and Nutrition,*

University of Nottingham School of Agriculture, Sutton Bonnington, Loughborough, Leicestershire, LE12 5RD, UK

A. TALLENTIRE, *Department of Pharmacy, University of Manchester, Manchester M13 9PL, UK*

R. H. TILBURY, *Tate and Lyle Ltd., Group Research and Development, PO Box 68, Reading, Berkshire RG2 6BX, UK*

B. J. TINDALL, *Department of Microbiology, The Medical School, University of Leicester, Leicester LE1 7RN, UK*

W. M. WAITES, *Agricultural Research Council, Food Research Institute, Colney Lane, Norwich NR4 7UA, UK*

M. WOODBINE, *Department of Applied Biochemistry and Nutrition, University of Nottingham, School of Agriculture, Sutton Bonnington. Loughborough, Leicestershire LE12 5RD, UK*

Preface

This volume is number 15 in the Technical Series of the Society for Applied Bacteriology. It contains papers that are based on contributions made at the Autumn Demonstration Meeting of the Society which was held at the Polytechnic of the South Bank in London on 18 October 1978.

The meeting aimed to bring together current research and modern techniques that are concerned, not with aspects of the growth of micro-organisms under optimal conditions, but rather with growth and survival under conditions that are far from optimal. Such conditions vary greatly in type, but have in common that they are widely regarded as imposing some sort of additional 'stress' on the microbial cell, and can all be regarded as more or less 'extreme'.

Extreme conditions covered by the various demonstrations included physical extremes, e.g. of heat, ionizing radiation, hyperbaric oxygen, reduced water activity and raised osmotic pressure. Other demonstrations were concerned with chemical and biochemical extremes, e.g. of low and high pH values, high concentrations of sulphur dioxide, nitrite, heavy metals, phenol, other disinfectants and chemicals, detergents and oils, as well as considering the somewhat neglected, but ecologically most important, environmental extreme represented by the very low nutrient concentrations in oligotrophic waters.

The organisms covered encompassed vegetative and spore forms of bacteria yeasts and moulds, and included organisms of public health significance as well as those involved in biodeterioration and those of ecological importance in the natural environment.

Thus, whilst not covering every possible environmental extreme to which micro-organisms have become adapted, the topics contributed to the meeting were sufficiently diverse to give a comprehensive overview, with some topics treated in considerable depth. This is reflected in the present volume, which contains papers based on the majority of the demonstrations that were shown.

We therefore thank all those who contributed to the meeting and who so promptly prepared the manuscripts on which this book is based. We also thank Mr Arnold Fox, and other members of the staff of the Department of Applied Biology at the South Bank Polytechnic for the hard work involved in hosting the meeting.

March 1980
G. W. GOULD
JANET E. L. CORRY

Contents

Colonization of Damp Organic Substrates and Spontaneous Heating 53

J. LACEY

The Growth of Microbes at Low pH Values . . . 71

M. H. BROWN, T. MAYES AND H. L. M. LELIEVELD

Growth in Hyperbaric Oxygen: Effect upon the Resistance to Antibacterial Agents and Morphology of *Pseudomonas aeruginosa* 99

M. A. KENWARD, S. R. ALCOCK AND M. R. W. BROWN

Microbial Growth and Survival in Oligotrophic Freshwater Environments

C. S. DOW AND A. LAWRENCE

Department of Biological Sciences, University of Warwick, Coventry, UK

The many microbial habitats and complex microbial communities found in the aqueous environment are continually changing in response to biotic and abiotic stress. The micro-organisms inhabiting such eco-systems are, consequently, under severe selective pressure to evolve means of coping with the changing nutrient and physiological conditions. This article is concerned with microbial adaptation to one of these stresses, low-level nutrient situations (i.e. oligotrophic freshwater systems) and concentrates primarily on the so-called budding and prosthecate bacteria and their responses to changes in nutrient status.

Micro-organisms possessing appendages which are an integral part of the cell have been collectively termed prosthecate bacteria, the term prostheca having been proposed by Staley (1968) and defined as follows: 'a semi-rigid appendage extending from a prokaryotic cell with a diameter which is always smaller than that of the mature cell and which is bounded by the cell wall'. Prosthecate bacteria can be sub-divided into two groups (Table 1): (i) those in which the prostheca(e) play an apparently obligate role in reproduction; (ii) those in which the prostheca(e) may be induced or repressed in response to environmental stimuli and/or play no part in the reproductive process.

These organisms are not characterized by a common physiological or nutritional property (Table 1) they are, however, indigenous to aqueous environments, particularly oligotrophic waters.

These bacteria seem to have come to terms with the nutrient variability and severity of an oligotrophic aqueous environment by the following means: (i) movement by flagella and gas vacuole production, i.e. chemo-tactic and phototactic responses; (ii) adhesion to surfaces; (iii) coloni-zation of favourable ecological niches; (iv) specialized dispersal phases; (v) formation of prosthecae.

TABLE 1

Sub-division of the prosthecate bacteria and their nutritional status

Prosthecate bacteria	Group (i) prostheca have an obligate role in reproduction	Group (ii) prostheca non-obligate/ not involved in reproduction	Nutritional status
Hyphomicrobium	+	−	aerobic heterotroph
Rhodomicrobium	+	−	photoheterotroph
Rhodopseudomonas palustris	+	−	photoheterotroph
Rhodopseudomonas viridis	+	−	photoheterotroph
*Caulobacter**	+	(+)	aerobic heterotroph
*Asticcacaulis**	+	(+)	aerobic heterotroph
Ancalomicrobium	−	+	aerobic heterotroph
Prosthecomicrobium†	−	+	aerobic heterotroph
Prosthecochloris†	−	+	obligate photoautotroph

*Caulobacter and *Asticcacaulis* form prostheca as part of the obligate temporal sequence of events occurring during the vegetative cell cycle. However, recent reports (Osley & Newton 1977) suggest that prostheca formation is not an obligate requirement for reproduction.

†Whether *Prosthecomicrobium* (Staley 1968) and *Prosthecochloris* (Gorlenko 1970) will suppress prostheca formation in situations of nutrient excess, in a similar manner to *Ancalomicrobium* (Whittenbury & Dow 1977) is not known.

The prosthecate bacteria, to varying degrees, incorporate all of these adaptations.

Enrichment and Isolation of Prosthecate Bacteria

Group (i) prosthecate bacteria (Table 1)

(prosthecae have an obligate role in reproduction)

This group of prosthecate bacteria present few problems with respect to enrichment and isolation, the procedures outlined below being simple and highly selective.

Hyphomicrobium

The selective pressures used to enrich hyphomicrobia from a wide range of habitats are their preference to grow on one-carbon compounds, particularly methanol, and their potential to use nitrate as an alternative to oxygen as terminal electron acceptor (Attwood & Harder 1972).

Medium. K_2HPO_4 1·74 g; $NaH_2PO_4.H_2O$ 1·38 g; $(NH_4)_2SO_4$ 0·5 g; $MgSO_4.7H_2O$ 0·2 g; $CaCl_2.2H_2O$ 0·025 mg; $FeCl_2.4H_2O$ 3·5 mg per litre of distilled water. This basic medium is supplemented with 0·5 ml of a trace element solution (Harder & Veldkamp 1967), KNO_3 (0·2 % w/v) and methanol (0·5 % v/v). The pH is adjusted to 7·0 with N NaOH.

Enrichments are carried out in liquid medium under an atmosphere of N_2 at 30°C in the dark (care must be taken to release the gas pressure which increases dramatically as a consequence of vigorous denitrification). Subsequent isolation and purification is performed on solid medium, lacking nitrate, under aerobic conditions at 30°C.

Rhodomicrobium *and* Rhodopseudomonas palustris
These photoheterotrophic bacteria can best be enriched by employing a malate salts medium of pH 5·5 for *Rhodomicrobium* and pH 7·0 for *Rhodopseudomonas palustris* (Whittenbury & Dow 1977). The basic medium consists of: NH_4Cl 0·5 g; $MgSO_4.7H_2O$ 0·4 g; $CaCl_2.2H_2O$ 0·05 g; NaCl 0·4 g; sodium hydrogen malate 1·5 g per litre of distilled water. The pH is adjusted with potassium hydroxide prior to auto-claving and 50 ml of 0·1 M phosphate buffer (of the desired pH) is added aseptically per litre. Incubation is in glass bottles, under N_2 at 30°C with an incident light intensity (tungsten lamps) of 1000 lx. Subsequent isolation and purification is by dilution of the enrichment cultures into agar deeps of the above medium. Alternatively, spread plates, incubated aerobically in the dark, may be used.

Group (ii) prosthecate bacteria (Table 1)

(prostheca not involved in reproduction and may or may not be non-obligate)

Multiappendaged prosthecate bacteria (e.g. Ancalomicrobium)
These organisms are the most difficult to enrich and isolate in pure culture *unless* they are severely stressed with nutrient limitation.
Enrichment. Oligotrophic freshwater samples are incubated static at room temperature for a period of 3–4 months—*no nutrient addition*. At the end of this time interval multiappendaged organisms should be observed by microscopy (preferably electron microscopy).
Isolation. Spread plates are prepared from the enrichment sample using both the pellicle and the body of the sample. The latter is important since many of the multiappendaged prokaryotes do not adhere to surfaces and are most abundant in this phase. Plates are incubated at room temperature over a period of a few weeks and as the colonies

appear they are patch plated on to fresh medium and examined by microscopy.

The major problem encountered with these organisms is that an increase in nutrient concentration will repress prostheca formation and morphologically the organism reverts to a rod-shape (this may not, of course, be true of all multiprosthecate bacteria).

Medium.   (a) filter sterilized pond water

           (b) filter sterilized pond water $+0.001\%$ (w/v) peptone.

The enrichment of *Caulobacter* and *Asticcacaulis* is essentially as described above, these organisms predominate in the pellicle after 2–3 weeks incubation. Isolation and purification is best facilitated on the following medium, incubation being at 30°C: peptone 2 g; yeast extract 1 g; $MgSO_4.7H_2O$ 0·2 g per litre of filtered pond water. A defined medium suitable for the growth of some caulobacters is: Na_2HPO_4 1·74 g; KH_2PO_4 1·06 g; NH_4Cl 0·5 g and 10 ml of concentrated base per litre of distilled water. After sterilization 0·2% (w/v) glucose is added as carbon and energy source.

Concentrated base: nitrilotriacetic acid 20 g; $MgSO_4.7H_2O$ 59·16 g; $CaCl_2.2H_2O$ 6·67 g; $(NH_4)_6Mo_7O.4H_2O$ 18·5 mg; $FeSO_4.7H_2O$ 198 mg; 'metals 44' 100 ml per litre of distilled water (the nitrilotriacetic acid is dissolved separately and neutralized with potassium hydroxide).

Metals 44: EDTA 2·5 g; $ZnSO_4.7H_2O$ 10·95 g; $FeSO_4.7H_2O$ 5·0 g; $MnSO_4.H_2O$ 1·54 g; $CuSO_4.5H_2O$ 392 mg; $Co(NO_3).6H_2O$ 250 mg; $Na_2B_4O_7.10H_2O$ 177 mg per litre distilled water (a few drops of concentrated H_2SO_4 added to retard precipitation) (Poindexter 1964).

Microscopy

Light micrographs were taken using high power ($\times100$) oil immersion phase optics. We have found that high contrast, low grain Kodak Panatomic x (32ASA) 35 mm film gives good reproduction (developed in Kodak D-19 developer (20°C) for 3 min).

To enhance the quality of light micrographs it is advisable to cast a uniform agar surface on a slide and inoculate this with the appropriate organism. This is done, as for a slide culture, by placing a few drops of molten agar onto a slide and applying a coverslip immediately. When the agar has set the coverslip is carefully removed leaving behind a flat uniform surface. Inoculation of such a surface reduces the number of organisms out of focus at any one time.

Electron micrographs were taken on an AEI Corinth 275. Sample preparation was either by negative staining or by casting a gold/palladium metal shadow (shadow angle of 65°) (Hayat 1972).

Functions of Prosthecae

What function(s) do the prosthecae serve? Although prosthecate bacteria, *Caulobacter* and *Hyphomicrobium* in particular, were first observed almost fifty years ago (Henrici & Johnson 1935) the question of prostheca function still remains unanswered. Suggested functions range from the prostheca acting as a "sucking proboscis" which enables the micro-organism (*Caulobacter* in this instance) to absorb nutrients from, and destroy, the host bacterium (Houwink 1951), to a floatation organelle to retard sedimentation and so maintain aerobic heterotrophs at the air-water interface (Poindexter 1964), and finally as structures which have evolved to enhance uptake by the cell of nutrients from low nutrient ecosystems. Several observations point to the last suggestion being the primary function; however, the consequences of having evolved prostheca(e) are far reaching, particularly with respect to the group (i) (see Table 1) prosthecate bacteria, all of which have assumed an obligate polar mode of growth. These consequences are discussed in detail later on.

It has long been realized that these particular micro-organisms are capable of growth and survival in extremely low nutrient aqueous environments (Henrici & Johnson 1935; Houwink 1952) and furthermore that prostheca length reflects the nutrient status of the ecosystem (Boltjes 1934). This variation in prostheca length in response to nutrient levels is most apparent with *Caulobacter* spp. grown in the presence and absence of phosphate (Schmidt & Stanier 1966), i.e. at different growth rates (Fig. 1). A similar correlation has been shown for *Hyphomicrobium* (Harder & Attwood 1978) (Table 2), *Rhodomicrobium vannielii* (Whittenbury & Dow 1977; Fig. 1) and an *Ancalomicrobium* isolate (Whittenbury & Dow 1977; Fig. 2).

The close correlation of prostheca length with nutrient concentration has prompted the suggestion that prosthecae serve to enhance nutrient

TABLE 2

Relationship between mother-cell size, mean prostheca length and growth rate in Hyphomicrobium X *grown in a methanol-limited continuous culture*

Growth rate (h^{-1})	Mean dimensions of mother cell (μm)	Mean prostheca length (μm)
0·02	0·7 ×0·4	3·9
0·05	0·9 ×0·6	2·6
0·11	1·0 ×0·6	1·2

Taken from Harder & Attwood (1978).

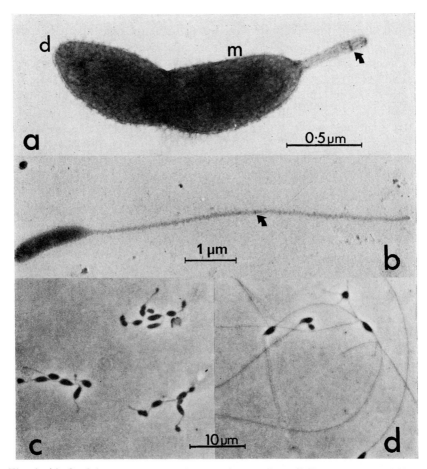

Fig. 1. (a) *Caulobacter* sp. grown in complex medium (150 mg peptone, 150 mg yeast extract, 250 mg glucose, 1 g NH_4Cl, 1 g $MgSO_4.7H_2O$, 0·5 g NaCl, 0·5 g $CaCl_2.6H_2O$, 0·1 M phosphate—pH 6·9 per litre distilled water) with a generation time of 2 h. d, daughter cell; m, mother cell. Uranyl acetate stained electron micrograph. (b) A *Caulobacter* sp. grown in dilute medium (100 mg peptone per litre distilled water, *no added phosphate*) with a generation time approximating to 10 h. Gold palladium shadowed electron micrograph. The crossbands characteristic of the *Caulobacter* prostheca are arrowed in both micrographs. (c) *R. vannielii* grown with a generation time of 6 h (medium containing 0·5 g NH_4Cl, 0·5 g $MgSO_4.7H_2O$, 0·05 g $CaCl_2.2H_2O$, 0·4 g NaCl, 1 g sodium hydrogen malate, 1 g sodium pyruvate, 0·1 M phosphate—pH 6·9, per litre distilled water). (d) *R. vannielii* grown with a generation time approaching 15 h (medium containing no added phosphate). Phase contrast photomicrographs.

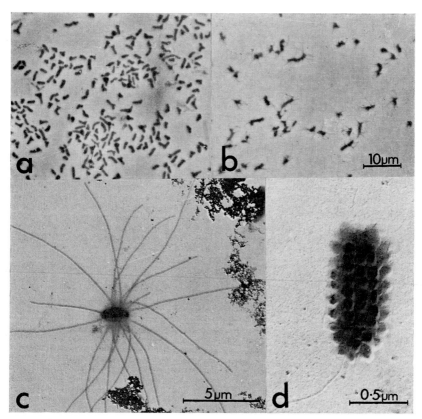

Fig. 2. (a) An *Ancalomicrobium* sp. grown in the presence of 100 μg ml⁻¹ peptone. (b) An *Ancalomicrobium* sp. grown in the presence of 1 μg ml⁻¹ peptone. Phase contrast photomicrographs. (c) Expression of prosthecae by an *Ancalomicrobium* sp. (d) A motile *Ancalomicrobium* cell in the repressed form. Gold palladium shadowed electron micrographs.

uptake by increasing the cell surface:volume ratio. To increase the cell's surface area, i.e. uptake area, long, thin integral cellular extensions (Fig. 3) are more effective than an increase in the surface area of the cell body proper as the latter quickly becomes subject to cell volume constraints. This concept has also been shown to be characteristic of non-prosthecate bacteria. Aerobic enrichment experiments using a chemostat support the view that diversity among bacteria in the natural environment is based, in part, on selection towards substrate concentration and that organisms showing a selective advantage at very low concentrations of limiting substrate appear to have a relatively high surface area to volume ratio (Kuenen *et al.* 1977).

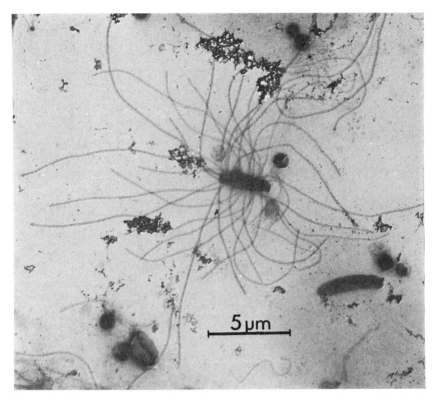

Fig. 3. A fully expressed prosthecate bacterium present in an oligotrophic freshwater lake sample. Gold palladium shadowed electron micrograph.

Physiological studies on isolated prosthecae from *Asticcacaulis*
biprosthecum (Porter & Pate 1975; Larson & Pate 1976) further sub-
stantiate the contention that some prosthecae, at least, represent specialized membrane systems for the adsorption and transport of nutrients.
The data obtained indicate that free glucose is accumulated within
prosthecae at a concentration 60–200 times above that present externally
and that these structures possess a respiration linked active transport
system for at least four amino acids, as well as for glucose.

The survival advantage conferred by possessing prosthecae is illustrated in experiments which monitor the changes in the microbial
population within a closed system over a lengthy period of time (Fig. 4).
In situ the heterotrophic prosthecate bacteria account for only a low
percentage of the total microbial population (Table 3), however, they
become the predominant species in a closed system deprived of additional
nutrients. It should be stressed that the majority of the prosthecate

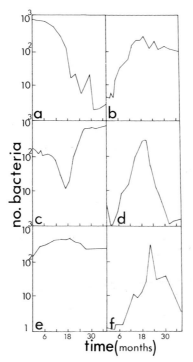

Fig. 4. Population graphs for a Draycote Water Reservoir sample followed over a three year period of incubation (static enrichment). In excess of 1000 cells were counted by electron microscopy and numbers plotted against time: (a) other bacteria; (b) multi-appendaged bacteria; (c) *Hyphomicrobium;* (d) *Planctomyces;* (e) *Caulobacter;* (f) gas-vacuolated rods.

bacteria are oligocarbophilic, i.e. they are capable of growth and reproduction in environments in which there is little or no detectable carbon substrate. Indeed it can be shown that trace amounts of C_1 and C_2 hydrocarbons in the atmosphere are sufficient to repress prostheca expression of the multiprosthecate species "enriched" in the closed system described above.

Physiology and Mode of Growth

In the majority of prosthecate bacteria a prerequisite to the evolving of prosthecae seems to be the adoption of a polar mode of growth, particularly evident with the group (i) (see Table 1) organisms. This has several major implications, which are: (i) the potential to form chains of cells or multicellular arrays, e.g. *Rhodomicrobium* and *Hyphomicrobium*

TABLE 3

Numbers of heterotrophic prosthecate bacteria in surface samples of freshwater lakes in Cumbria

Sample site	Viable count/ml*	Prosthecate bacteria/ml (heterotrophs)†	%	Trophism
Ennerdale water	$7\cdot2 \times 10^3$	$8\cdot8 \times 10^2$	12·2	oligotrophic
Thirlmere	$9\cdot8 \times 10^3$	$8\cdot9 \times 10^2$	9·08	intermediate
Ullswater	$5\cdot8 \times 10^4$	$3\cdot9 \times 10^3$	6·7	intermediate
Haweswater	$1\cdot6 \times 10^4$	$7\cdot4 \times 10^2$	6·2	oligotrophic
Wise Een Tarn	$4\cdot3 \times 10^3$	$6\cdot9 \times 10^2$	16·0	oligotrophic
Three Dubbs Tarn	$9\cdot7 \times 10^2$	$2\cdot3 \times 10^2$	23·7	oligotrophic
Moss Eccles Tarn	$5\cdot8 \times 10^3$	$7\cdot8 \times 10^2$	13·4	oligotrophic
Windermere North	$8\cdot2 \times 10^3$	1×10^2	1·2	eutrophic
Windermere South	$9\cdot0 \times 10^3$	$3\cdot5 \times 10^2$	3·9	becoming eutrophic
Esthwaite	$1\cdot4 \times 10^4$	2×10^2	1·4	eutrophic
Coniston	$9\cdot3 \times 10^3$	1×10^2	1·1	eutrophic

*Viable counts were obtained using the following medium: 150 mg peptone, 150 mg yeast extract, 250 mg $(NH_4)_2SO_4$, 1 g glucose, 0·1M phosphate buffer pH 6·8, 10 ml vitamin solution (Staley 1968), 1·5% (w/v) Bacto agar to 1 litre distilled water.

†Plates were incubated in the dark at 30°C for four weeks, counted and each colony type examined microscopically.

(Fig. 5); (ii) marked cellular orientation. The cell is polarized having a distinct "front" and "back" end. Division is asymmetric (Fig. 1a); (iii) asymmetric division generates the potential to evolve dimorphic (e.g. *Caulobacter*) and polymorphic (e.g. *Rhodomicrobium*) vegetative cell cycles.

Many species of micro-organisms have evolved mechanisms by which they can form colonies in the natural environment. These adaptations range from holdfast formation, e.g. *Caulobacter*, *Planctomyces*, *Hyphomicrobium*, retention in a sheath (e.g. *Sphaerotilus natans*) or envelope (e.g. *Thiocystis* and *Thiocapsa*), secretion of a polysaccharide matrix, and progeny cells remaining attached so generating multicellular complexes (e.g. *Rhodomicrobium*, *Anabaena*). Colony formation may be considered to impart the following advantages:

    (i) Nutrients aggregate at interfaces; consequently, any organisms which can attach themselves to the substratum and so remain localized in the proximity of the interface will have a selective advantage. Hence the solid-liquid interface is colonized by holdfast formation or multicellular arrays such as *Rhodomicrobium* and *Hyphomicrobium* (Fig. 5). In this context the phenotypic expres-

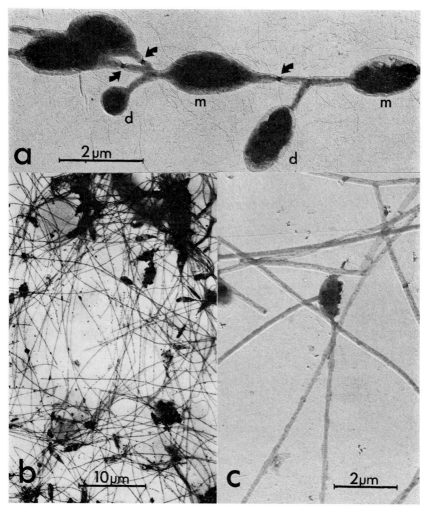

Fig. 5. (a) Gold palladium shadowed electron micrograph of a *Rhodomicrobium* multicellular array. The arrows indicate the plugs formed within the prostheca at division. The continuity of mother cell (m) and daughter cell (d) or bud is evident. (b & c) *Hyphomicrobium* multicellular matrix expressed in oligotrophic lake water. Gold palladium shadowed electron micrographs.

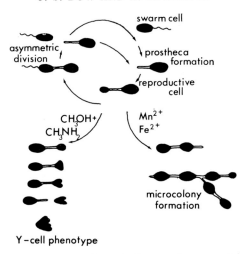

Fig. 6. Diagrammatic representation of phenotypic expression in *Hyphomicrobium* sp.

sion of *Hyphomicrobium* is of interest (Fig. 6). In the presence of Mn^{2+} or Fe^{2+} this organism becomes encrusted in a hydroxide deposit (Hirsch 1968; Tyler 1970), a characteristic which led to the suggestion that the polar mode of growth was essential to prevent total encrustation of the organism, i.e. new growth is always away from the deposit. The ability of *Hyphomicrobium* to colonize and exploit the solid-liquid interface is best illustrated by its presence in Tasmanian hydroelectric pipelines carrying water high in Mn^{2+} (Tyler & Marshall 1967).

The multiappendaged prosthecate bacteria (Figs 2 & 3) are mostly aerobic heterotrophs which do not adhere to the substratum. The possession of multiple prosthecae therefore may be an alternative to localization at an interface in that a greater adsorption area is presented to the environment. This group of prosthecates, with the exception of *Prosthecochloris*, are obligate aerobes. Consequently they predominate at the air-liquid interface and it has been shown that in at least one species of *Prosthecomicrobium* gas vacuoles are produced (Walsby 1975) to maintain the organisms at the desired interface.

(ii) Colonization presents an arena for species interaction. The formation of polysaccharide matrices is a well-known phenomenon in the aquatic environment (Fig. 13) and apart from the immediate advantages conferred by the polysaccharide (nutrient adsorption, resistance to desiccation, etc.) on the producing organism it presents a matrix for attachment of other micro-organisms, a prerequisite to microbial interaction in an aquatic environment.

(iii) Where multicellular units are produced, e.g. *Rhodomicrobium* and *Anabaena*, differentiation of some of the cells in the complex increases the survival potential of the colonial morphology over that which could be expressed by a single cell of the same species. This is clearly the case with *Anabaena* (Fig. 7), however, whether *Rhodomicrobium* behaves in an analogous fashion with differentiation of some of the cells in the complex to serve a specific physiological function which benefits the colony as a whole has not been shown.

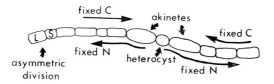

Fig. 7. Diagrammatic representation of cell differentiation in an *Anabaena* filament, the advantages possessed by such a multicellular system are: akinete—resistant cell, i.e. survival; heterocyst—N_2 fixation; vegetative cell—CO_2 fixation. The filament is also motile.

Dispersal and Swarm Cells

A consequence of a colonial habit, or of attachment in general, is the requirement for a dispersal phase, essential for species survival in an unstable ecosystem. All of the group (i) (see Table 1) prosthecate bacteria have a motile cell expressed during the vegetative cell cycle, indeed the cell cycles of each are remarkably similar and may be generalized as in Fig. 8. That these swarm cells are a specialized dispersal phase is supported by recent observations on the analysis of the cell type distribution during growth of *Hyphomicrobium*, *Caulobacter* and *Rhodomicrobium*. Cell type distribution analysis of such complex heterogeneous

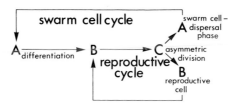

Fig. 8. Generalized cell cycle of the group (i) prosthecate bacteria described in Table 1.

cell populations (Fig. 9) is possible using an electronic particle counter linked to a cell volume distribution analyser (Coulter Electronic's ZBI and C1000 system; Fig. 10). The relative numbers of each cell type can be obtained from the cell volume distribution profiles and the ratios of one cell type to another plotted against time (Fig. 11). These data indicate that during the exponential phase of batch growth the cell types making up the population remain constant. With the onset of the stationary phase, however, the number of motile swarm cells increases dramatically. This can be interpreted as follows. When the system becomes limited for an essential nutrient, swarm cell production continues as long as is possible; however, the swarm cells produced are inhibited from initiating the differentiation sequence since, like many biological differentiation systems, once the sequence is initiated it must run to conclusion. Consequently it would be detrimental to species survival not to have an inhibition switch linked to the nutrient status of the ecosystem, i.e. the

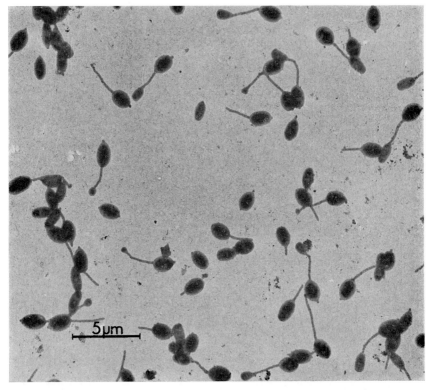

Fig. 9. Gold palladium shadowed electron micrograph of a heterogeneous *Hyphomicrobium* population (medium prepared after Attwood & Harder 1972).

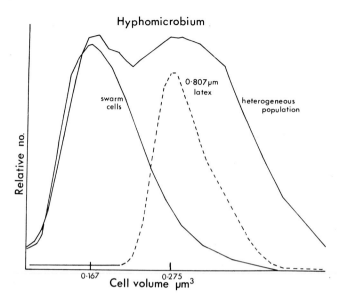

Fig. 10. Cell volume distribution profiles (Dow 1976) of a heterogeneous *Hyphomicrobium* population, a homogeneous *Hyphomicrobium* swarm cell population and calibration latex having a spherical diameter of 0·807 μm.

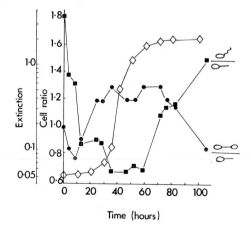

Fig. 11. Ratios of the relative numbers of *Hyphomicrobium* plotted against time. Swarm cells/stalked cells = ■; reproducing cells/stalked cells = ●; △, growth curve.

swarm cell only initiates differentiation and growth in a favourable environmental situation, an essential requirement for a dispersal phase. A similar response to the nutrient status of the environment can be demonstrated for *Caulobacter* and *Rhodomicrobium* (Fig. 12).

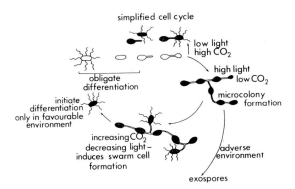

Fig. 12. Diagrammatic representation of environmentally induced differentiation in *R. vannielii*.

Bacteria with Acellular Extensions

In addition to the prosthecate bacteria, electron microscope studies of oligotrophic freshwater environments indicate the prevalence of bacteria with acellular extensions (Fig. 13). One of the most interesting of these is *Planctomyces* the cell cycle of which bears considerable resemblance to that of *Caulobacter* (Fig. 14). The *Planctomyces* stalk is however, acellular (Fig. 15). Another unusual characteristic of this organism is the formation of an extensive array of fimbriae (Fig. 15). It may be argued that lacking, a prostheca, *Planctomyces* has evolved the fimbriae as an alternative nutrient adsorption system. This may be less efficient than prosthecae since in the closed system studies (Fig. 3) *Plantomyces* competes efficiently initially but subsequently declines.

The remarkable likeness of this organism to *Caulobacter*, the only major distinction being the acellular nature of the stalk, raises the question as to the function of this structure in *Planctomyces*. It is difficult to envisage any advantage being conferred other than elevation above the substratum, by doing so the organism would generate a larger "sphere of influence", an argument which would hold equally for the *Caulobacter* prostheca.

Fig. 13. (a) Fimbriae-like extensions on a bacterium observed in an oligotrophic "enrichment". (b) Bacterial spines (also observed by Easterbrook *et al.* 1973). (c and d) Fimbriae. (e and f) Polysaccharide excretions. (g) Acellular extensions of unknown composition. (h) Polar tufts of flagella (also observed by Strength & King 1971). Gold palladium shadowed electron micrographs.

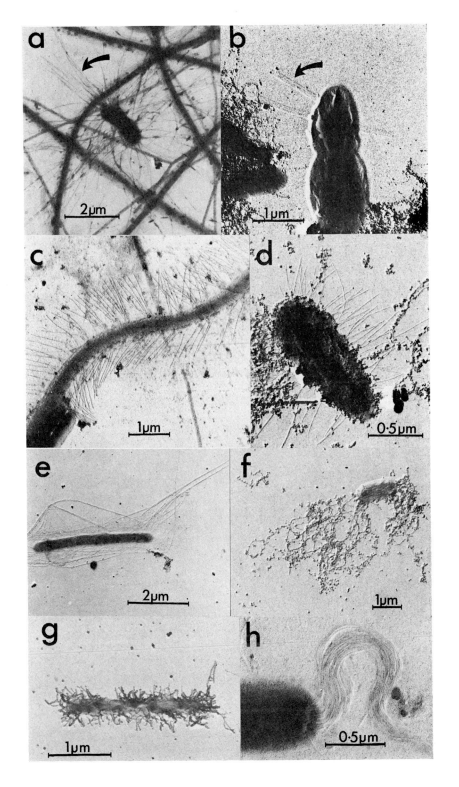

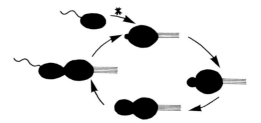

Fig. 14. Diagrammatic representation of the *Planctomyces* cell cycle: *, the relative order of Fimbriae, stalk and holdfast formation is not known.

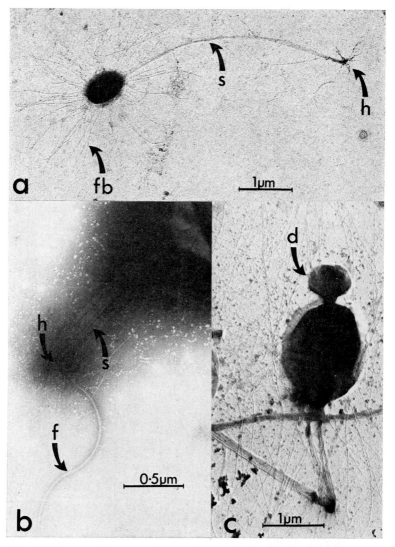

Fig. 15. (a) Gold palladium shadowed electron micrograph of *Planctomyces*. (b) Phosphotungstic acid-stained electron micrograph showing the fibrillar nature of the *Planctomyces* stalk. The *Planctomyces* flagellum is also apparent. (c) Gold palladium shadowed electron micrograph showing *Planctomyces* reproduction. s, stalk; h, holdfast; fb, fimbriae; f, flagellum; d, daughter cell.

Conclusions

In conclusion, the prosthecate bacteria form a group of micro-organisms which seem to have become highly adapted to growth and survival in an extreme environment—that of oligotrophic freshwater. With the exception of prosthecae these adaptations are not unique to prosthecate bacteria but as a group they are perhaps the most highly adapted pro-karyotes in the aquatic environment.

References

ATTWOOD, M. & HARDER, W. 1972 A rapid and specific enrichment procedure for *Hyphomicrobium* sp. *Antonie van Leeuwenhoek* **38**, 369–378.

BOLTJES, T. 1934 Onderzoekingen over Nitrificeerende Bacterien. *Thesis*, Techn. Hoogeschool, Delft.

DOW, C. S. 1976 Cell volume analysis. *2nd International Symposium on Rapid Methods and Automation in Microbiology*, pp. 97–98. Publications of Learned Information (Europe) Ltd.

EASTERBROOK, K. B., MCGREGOR-SHAW, J. B. & MCBRIDE, R. P. 1973 Ultra-structure of bacterial spines. *Canadian Journal of Microbiology* **19**, 995–997.

GORLENKO, V. M. 1970 A new phototrophic green sulphur bacterium *Prosthecochloris aestuarii* nov. gen. nov. spec. *Zeitschrift für Mikrobiologie* 147–149.

HARDER, W. & ATTWOOD, M. M. 1978 Biology, physiology and biochemistry of Hyphomicrobia. *Advances in Microbial Physiology* **16**, 303–356.

HARDER, W. & VELDKAMP, H. 1967 A continuous culture study of an obligately psychrophilic *Pseudomonas* species. *Archiv für Mikrobiologie* **59**, 123–130.

HAYAT, M. A. 1972 *Principles and Techniques of Electron Microscopy* Vol. II, New York & London: Van Nostrand Reinhold Company.

HENRICI, A. T. & JOHNSON, D. E. 1935 Studies on freshwater bacteria II. Stalked bacteria; a new order of Schizomycetes. *Journal of Bacteriology* **30**, 61–93.

HIRSCH, P. 1968 Biology of budding bacteria IV. Epicellular deposition of iron by aquatic budding bacteria. *Archiv für Mikrobiologie* **60**, 201–216.

HOUWINK, A. L. 1951 *Caulobacter* versus *Bacillus* spec. *Nature* **168**, 654.

HOUWINK, A. L. 1952 Contamination of electron microscope preparations. *Experientia* **8**, 385.

KUENEN, J. G., BOONSTRA, J., SCHRODER, H. G. J. & VELDKAMP, H. 1977 Competition for inorganic substrates among chemo-organotrophic and chemo-lithotrophic bacteria. *Microbial Ecology* **3**, 119–130.

LARSON, R. J. & PATE, J. L. 1976 Glucose transport in isolated prosthecae of *Asticcacaulis biprosthecum*. *Journal of Bacteriology* **126**, 282–293.

OSLEY, M. A. & NEWTON, A. 1977 Mutational analysis of developmental control in *Caulobacter crescentus*. *Proceedings of the National Academy of Sciences USA* **74**, 124–128.

POINDEXTER, J. S. 1964 Biological properties and classification of the *Caulobacter* group. *Bacteriological Reviews* **28**, 231–295.

PORTER, J. S. & PATE, J. L. 1975 Prosthecae of *Asticcacaulis biprosthecum*: a system for the study of membrane transport. *Journal of Bacteriology* **122**, 976–986.

SCHMIDT, J. M. & STANIER, R. Y. 1966 The development of cellular stalks in bacteria. *Journal of Cell Biology* **28**, 423–436.

STALEY, J. T. 1968 *Prosthecomicrobium* and *Ancalomicrobium* new prosthecate freshwater bacteria. *Journal of Bacteriology* **95**, 1921–1942.

STRENGTH, W. J. & KING, N. R. 1971 Flagellar activity in an aquatic bacterium. *Canadian Journal of Microbiology* **17**, 1133–1137.

TYLER, P. A. 1970 Hyphomicrobia and the oxidation of manganese in aquatic ecosystems. *Antonie van Leeuwenhoek* **36**, 567–578.

TYLER, P. A. & MARSHALL, K. C. 1967 Pleomorphy in stalked budding bacteria. *Journal of Bacteriology* **93**, 1132–1136.

WALSBY, A. E. 1975 Gas vesicles. *Annual Review of Plant Physiology* **26**, 427–439.

WHITTENBURY, R. & DOW, C. S. 1977 Morphogenesis and differentiation in *Rhodomicrobium vannielii* and other budding and prosthecate bacteria. *Bacteriological Reviews* **41**, 754–808.

Monitoring the Effects of 'Acid Rain' and Sulphur Dioxide upon Soil Micro-organisms

P. INESON AND T. R. G. GRAY

Department of Biology, University of Essex, Wivenhoe Park, Colchester, UK

In the UK only a small proportion of the airborne sulphur compounds have natural origins, the predominant form being sulphur dioxide anthropogenically generated as a consequence of fuel combustion. Natural levels of sulphur dioxide in the atmosphere are of the order of $3 \ \mu g \ m^{-3}$, yet mean annual concentrations around urban and industrial regions may be of the order of $200 \ \mu g \ m^{-3}$ (D.O.E. 1976).

Sulphur dioxide can affect the growth of higher plants directly, at concentrations occurring in ambient air (Bell & Clough 1973). There is a possibility that this gas may affect plant growth indirectly via action on soil micro-organisms and nutrient transformations for which they are responsible.

The antiseptic properties of SO_2 have been realized for many years and have been the basis for its use in the sulphuring of wines for at least 500 years (Fox & Cameron 1961). The effects upon micro-organisms at low concentrations are evidenced by the absence of lichens around industrial/urban areas (Gilbert 1970), and by the observed relationship between SO_2 concentrations and the incidence of such organisms as *Rhytisma acerinum, Hysterium pulicare* and *Diplocarpon rosae* (Saunders 1966; Skye 1968; Bevan & Greenhalgh 1976). The effects of SO_2 upon plant pathogenic micro-organisms have been reviewed by Heagle (1973) and the effect of air pollutants upon micro-organisms has been extensively considered by Babich & Stotzky (1974). There is virtually no information about the effect of SO_2 upon soil micro-organisms.

Oxidation of sulphur dioxide in the atmosphere can result in the formation of sulphate ions in rainfall, which may depress the pH. For example, the mean pH of rainfall in certain areas of Scandinavia is said to have decreased from a pH $>6 \cdot 0$ in 1956 to a pH $4 \cdot 0$–$4 \cdot 5$ in 1966, as a consequence of an increased 'burden' of sulphate from the more industrialized countries of Europe (Brosset 1973). Similarly, concern has been expressed in the United States about the extent and implications of 'acid rain' (US Department of Agriculture and the Ohio State University 1975). Lime applications to agricultural soils are sufficient to counteract the effects of increased sulphur deposition, yet natural and forest soils receiving no such inputs are at risk. Changes in the structure of soils' microbial populations may occur with resulting reductions in the decomposition of plant litter and nitrogen mineralization rates (Odén 1968; D.O.E. 1976).

The present paper is concerned with some of the techniques we have been using to determine the impact that sulphur dioxide and 'acid rain' have upon soil micro-organisms.

Use of Infra-red Gas Analysis for Measuring Soil Respiration

Carbon dioxide evolution is a widely used index of metabolic activity of the decomposer population. Many techniques have been used to determine rates of soil respiration, ranging from simple inverted boxes containing absorbing reagents (Witkamp 1966) to quite sophisticated microcosms (Bond et al. 1976).

Carbon dioxide is particularly suitable for determination in low concentrations by means of infra-red gas analysis, because it absorbs radiation very intensely in the infra-red region of the spectrum. Infra-red gas analysers can specifically measure the quantity of CO_2 in a gas mixture, offering the advantages of being continuous, accurate and non-destructive. Unlike many other of the respirometric techniques output is suitable for channelling to a recorder.

Gas to be analysed is passed through an optical cell situated in the path between a source of low energy infra-red radiation and a detector. Radiation in a particular band is absorbed by the CO_2 and reduces the level of energy reaching the detector. This change of energy is amplified to give the analyser output signal. A detailed description of the basis of infra-red gas analysis is given by Janac et al. (1971).

A useful analyser for such measurements is that supplied by Analytical Development Co. Ltd., Hoddesdon (type 225/2). This is a non-dispersive double beam instrument, which is constructed to measure differential increases in CO_2 concentration up to 50 parts/10^6. The analyser is supplied by a six-channel gas handler (Type WA161, Analytical Development Co. Ltd.), fitted with a sequential timer circuit which routes gas either through the analyser or to waste during a six minute cycle. Air flow through the analyser may be varied between 0–2·0 l min $^{-1}$ depending upon the intended application.

The gas handler draws air from seven respiration chambers, one of which remains empty and acts as a reference channel. Air is supplied to the chambers from a common supply which has been forced, under pressure, through a humidifier (see Fig. 1). Prior to entering the chambers excess pressure is dissipated and the gas handler pump becomes respons-

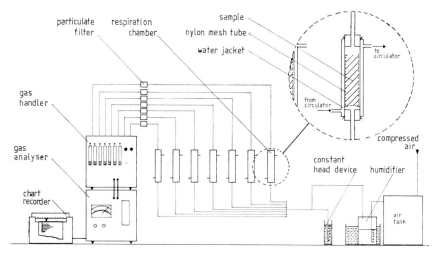

Fig. 1. Infra-red gas analysis apparatus used to measure soil respiration.

ible for air flow. After leaving the respiration chambers the gas is filtered to remove particulates and is routed to the analyser, or to waste, according to the six minute cycle. The output signal from the analyser is fed to a suitable chart recorder.

The respiration chambers are of glass construction (26 cm long, 3 cm internal diam.) having a water jacket for temperature control. The water jacket is fed by a thermocirculator (Churchill Instrument Company, Middlesex, England) from a water bath reservoir, which also houses the humidifier. The working temperature range is $-5°$ to $30°C$.

Measurement of Total Respiration

Soil or litter from the field is placed in a 1 mm mesh nylon tube (20 cm long, 25 cm wide) and allowed to equilibrate at the temperature of analysis (usually $20°C$). The tubes are then placed, entire, into the respiration chambers and measurements of carbon dioxide evolution made over a specified period. Six samples may be analysed simultaneously and, after assessment of respiration, the dry weight of the sample under study determined.

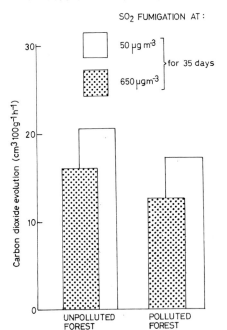

Fig. 2. Effect of sulphur dioxide upon respiration of *Pinus nigra* litter from two forests.

Figure 2 showed the effect of exposing *Pinus nigra* litter derived from two contrasting forest sites to sulphur dioxide gas. The litter was exposed at concentrations of 50 and 650 μg m^{-3} for 35 days, and respiration rate subsequently determined at 20°C, with a gas flow rate of 0·8 l min^{-1}. A highly significant reduction in the respiration rate of both litters occurred as a consequence of fumigation at 650 μg m^{-3}. Although such concentrations were common in urban and industrial regions at the turn of the century, by today's standards they are considered extreme. Similar reductions in respiration rate occurred in litters derived from polluted and non-polluted forests.

Partitioning of Respiration

If soil is acidified by airborne sulphur pollutants, fungi and acidophilic bacteria may be favoured at the expense of neutrophilic bacteria. This might be detected by determining the bacterial and fungal contributions to soil respiration, using the selective inhibition technique of Anderson & Domsch (1975). This technique involves the rapid and selective suppression of fungi and bacteria, experiments being performed over short time intervals using glucose amended soils. Measurements of changes in carbon dioxide evolution enable one to determine the relative metabolic activity of these two groups. Differentiation is based upon the selective inhibition of protein synthesis in prokaryotic and eukaryotic cells by the antibiotics, streptomycin and actidione. The dose rates for glucose and the antibiotics are determined by a series of preliminary experiments (Anderson & Domsch 1973). In the case of *P. nigra* litter layer material, glucose is applied at 400 parts 10^{-6}, streptomycin and actidione being applied at 200 parts 10^{-6}.

Litter is taken from the field, brought to saturation point with distilled water, and then placed inside 5 cm diameter plastic tubing. The tube is then sliced on a wood mitre, being cut so as to enable passage of the litter through a 4 mm sieve. At this stage a sub-sample is removed for rapid dry weight determination using an infra-red oven. The remainder is placed into a plastic bag and mixed with the glucose addition. Glucose is added as a powder, mixing being achieved by attaching the plastic bag to a flask shaker (Gallenkamp & Co. Ltd., London) for five minutes at maximum speed. After mixing, the sample is divided into four aliquots each receiving one of the following treatments: actidione; streptomycin; streptomycin+actidione; no addition. Again, additions are made as

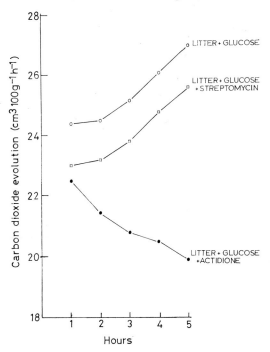

Fig. 3. Partitioning of litter respiration of *Pinus nigra*.

powders, talcum powder being used as a diluent, and mixing is carried out as outlined above. Samples are then weighed into nylon mesh bags and placed into the respiration chambers. Carbon dioxide evolution is monitored over a maximum of six hours.

Figure 3 shows the results of such an experiment using *P. nigra* litter layer material. On this occasion relative contributions of fungi and bacteria are respectively 80 : 20. Results from simulated acid rain work currently in progress suggest that acidification does not cause any major shift in the relative importance of these two groups.

References

ANDERSON, J. P. E. & DOMSCH, K. H. 1973 Quantification of bacterial and fungal contributions to soil respiration. *Archives für Mikrobiologie* **93**, 113–127.

ANDERSON, J. P. E. & DOMSCH, K. H. 1975 Measurement of bacterial and fungal contributions to respiration of selected agricultural and forest soils. *Canadian Journal of Microbiology* **21**, 314–322.

BABICH, H. & STOTZKY, G. 1974 Air pollution and microbial ecology. *Critical Reviews in Environmental Control* **4**, 353–421.

BELL, J. N. B. & CLOUGH, W. S. 1973 Depression of yield in rye grass exposed to sulphur dioxide. *Nature, London* **241**, 47–49.

BEVAN, R. J. & GREENHALGH, G. N. 1976 *Rhytisma acerinum* as a biological indicator of pollution. *Environmental Pollution* **10**, 271–285.

BOND, H., LIGHTHART, B., SHIMABUKU, R. & RUSSELL, L. 1976 Some effects of cadmium on coniferous forest soil and litter microcosms. *Soil Science* **121**, 278–287.

BROSSET, C. 1973 Airborne acid. *Ambio* **2**, 2–9.

D.O.E. 1976 Effects of airborne sulphur compounds on forests and freshwaters. *Pollution Paper No.* 7, London: H.M.S.O.

FOX, B. A. & CAMERON, A. G. 1961 *A Chemical Approach to Food and Nutrition* London: University of London Press Ltd.

GILBERT, O. L. 1970 A biological scale for the estimation of sulphur dioxide pollution. *New Phytologist* **69**, 627–630.

HEAGLE, A. S. 1973 Interactions between air pollutants and plant parasites. *Annual Review of Phytopathology* **11**, 365–385.

JANAC, J., CATSKY, J. & JARVIS, P. G. 1971 Infra-red gas analysers and other physical analysers. In *Plant Photosynthetic Production Manual of Methods* eds Sestak, Z., Catsky, J. & Jarvis, P. G. pp. 111–193. The Hague: Junk.

ODEN, S. 1968 Nederbörders och luftens försuring dess anaker, förlopp och verkani olika miljöer. *Swedish Ecology Committee Bulletin* No. 1.

SAUNDERS, P. J. W. 1966 The toxicity of sulphur dioxide to *Diplocarpon rosae* Wolf causing blackspot of roses. *Annals of Applied Biology* **58**, 103–114.

SKYE, E. 1968 Lichens and air pollution. *Acta Phytogeographica Suecica* **52**, 1–123.

UNITED STATES DEPARTMENT OF AGRICULTURE & THE OHIO STATE UNIVERSITY 1975 *Proceedings of the First International Symposium on Acid Precipitation and the Forest Ecosystem* Columbus, Ohio: Ohio State University.

WITKAMP, M. 1966 Decomposition of leaf litter in relation to environment, microflora and microbial respiration. *Ecology* **47**, 194–201.

The Isolation of Alkalophilic Bacteria

W. D. Grant and B. J. Tindall

Department of Microbiology, University of Leicester, Leicester, UK

Over the last few decades, many reports have appeared of organisms growing at extreme pH values. Most of the reports however, concern acidic environments, and little information exists about the alkaline limits of life, testified to by the paucity of detail in review articles concerned with the effects of pH (Vallentyne 1963; Skinner 1968; Kushner 1971; Langworthy 1978).

There are a number of man-made highly alkaline environments, e.g. food processing waste, textile wastes, but naturally occurring high alkaline ($>$pH 10) environments are relatively uncommon. There are certain desert soils where pH values of 10 and above have been recorded (Rupela & Tauro 1973) but the most alkaline, naturally occurring environments on earth are the soda lakes and soda deserts, now restricted to a few areas (for a list see Te-Pang 1890) where pH values of 10 and above are common. The areas where such lakes and deserts are to be found are mostly closed basins where evaporative concentration takes place. The genesis of the alkalinity is open to some dispute, but the most widely accepted explanation is that a weak solution of metal bicarbonates, largely sodium, leached from rocks by CO_2-charged surface waters, concentrates in the absence of significant amounts of calcium (which removes alkalinity by calcite precipitation) to produce a $NaHCO_3/Na_2CO_3$ brine (Hardie & Eugster 1970). At pH values above 10, Na_2CO_3 preponderates (Golterman & Clymo 1971), and most of the soda lakes have from 2–10% (w/v) Na_2CO_3 at pH values of 10–11, with varying amounts of other salts, often including NaCl as a major component. Extreme examples such as the Wadi Natrun in Egypt and Lake Magadi in Kenya have crystalline deposits reflecting total dissolved solids of up to 40% (w/v). Dilute alkaline springs are also found where the calcium content may be high (Souza *et al.* 1974)—in these cases the genesis of the alkalinity is not clear, and obviously different from that in the soda lakes.

No definition of the term alkalophile seems to exist (more correctly alkaliphile from the Arabic root, but the term alkalophile is now in common usage and will continue to be used in this article). However, the assumptions generally implicit in its use appear to be that those organisms incapable of growing at neutrality, capable of growth at pH 10 and above, and with pH optima at pH 9 or above are designated alkalophiles. The much larger group of organisms capable of growing over a wider range of pH from the acid side of neutrality to pH values in excess of 10, and with pH optima closer to neutrality should therefore be designated alkalotolerant.

Eukaryotic micro-organisms have been described in alkaline habitats (Souza *et al.* 1974; Siegel & Giumarro 1965; Langworthy 1978) but pure culture studies have seldom been made, so it is not clear if these organisms are alkalophiles or merely alkalotolerant. A considerable number of alkalotolerant species of bacteria are known from a wide variety of habitats, including members of the genera *Streptococcus* (Chesbro & Evans 1959), *Agrobacterium* (Allen & Allen 1950) and *Bacillus* (Bornside & Kallio 1956; Chislett & Kushner 1961). Alkalophilic species of bacteria have been isolated from a variety of habitats, some not particularly alkaline in nature. Most of the strains have been soil isolates of Gram positive sporing rods assigned to the genus *Bacillus*. Probably the best known alkalophile is *B. pasteurii*, a common soil isolate, which requires both high pH and NH_3 (Wiley & Stokes 1962). Other soil isolates include *B. alcalophilus* (Boyer & Ingle 1972; Vedder 1934) and the large number of incompletely characterized *Bacillus* spp. described by Horikoshi and his coworkers (Horikoshi 1971*a,b*; 1972; Yamamoto *et al.* 1972; Nakamura & Horikoshi 1976). Horikoshi has also isolated a *Bacillus* sp. from rayon waste (Ikura & Horikoshi 1977), and Ohta and his colleagues have isolated a *Bacillus* sp. from fermenting indigo leaves (Ohta *et al.* 1975). The extraordinary organism *Kakabekia barghoorniana* described by Siegel (Siegel & Siegel 1970; Siegel 1977) has been isolated from several soil samples. This particular organism grows in very high concentrations of NH_4OH, but has never been grown in the absence of soil particles. Alkaline springs have yielded several alkalophilic bacteria, including a coryneform organism (Souza & Deal 1977), a *Flavobacterium* sp., and an unidentified anaerobe (Souza *et al.* 1974; Deal *et al.* 1975).

Potentially the best source of alkalophilic micro-organisms, the soda lakes and soda deserts, are virtually unexplored microbiologically. No consideration of these environments would be complete without some mention of the massive blooms of blue green algae which occur in soda lakes periodically at pH values considerably in excess of 10. The most frequently described species are members of the genera *Spirulina* and

Anabaenopsis (Talling & Talling 1965; Melack & Kilham 1974). These algae have seldom if ever been isolated from these environments and studied in pure culture, so it is not clear if these organisms are true alkalophiles. However, in general there is evidence that blue green algae are more abundant in alkaline waters (Fogg 1956) and in some cases may be true alkalophiles. Bacteriological studies on soda lakes are scarce. Some years ago Isachenko (1951) observed several morphological types of bacteria in Crimean soda lakes, some of which were photosynthetic bacteria. Reports of pure culture studies on isolates from such lakes have appeared only recently and so far are restricted to photosynthetic bacteria. Imhoff and Trüper (1977), and Grant *et al.* (1979) have described alkalophilic *Ectothiorhodospira* spp. from different soda lakes. It is to be expected that these lakes will yield a wide variety of different types of alkalophilic bacteria in the near future.

The biochemical basis of alkalophily remains to be elucidated. Ohta *et al.* (1975) and Koyama *et al.* (1976) have shown that membrane exclusion mechanisms may be important in allowing the transport of charged compounds. Recently the problems posed by alkalophily in relation to ATP synthesis have been explored in *B. alcalophilus* (Guffanti *et al.* 1978) and alkalophiles may prove to be useful tools in furthering our understanding of the mechanics involved in energy generation.

Enrichment and Isolation Procedures

In principle, there is no reason why a variety of physiological groups of bacteria should not be isolated from an appropriate environment. In most cases the pH of the initial enrichment is achieved and maintained by adding an appropriate quantity of Na_2CO_3 (or $Na_2CO_3.10H_2O$). Occasionally NaOH has been used (Souza *et al.* 1974). Na_2CO_3 is the additive of choice for many enrichments in view of the high concentrations of the compound to be found in most naturally occurring alkaline environments, and there is some evidence that growth is enhanced in its presence (Nakamura & Horikoshi, 1976; Imhoff & Trüper 1977; Grant *et al.* 1979). pH values of 10–11 can be achieved in most media by the addition of 1–5% (w/v) Na_2CO_3 or the equivalent amount of $Na_2CO_3.10H_2O$. The pH of the enrichments should be checked and adjusted if incubation periods longer than 48 h are to be used. High pH media quickly absorb CO_2 from the atmosphere and drop markedly in pH after a relatively short period. For this reason growth after some time in media with a starting pH in excess of 11 should be checked very carefully, and claims of growth at pH 12–13 (Allen & Allen 1950; Kingsbury 1954) are to be viewed with suspicion unless careful pH measurements have been made. At this present time the highest pH recorded for growth in pure culture may be as low as 11·5.

It is not possible to maintain pH values in excess of 11·0 with the Na_2CO_3 system due to the buffering of atmospheric CO_2, but if a constant pH monitoring system is available then automatic addition of NaOH can be used to retain very high pH values (Souza et al. 1974; Souza & Deal 1977). However, in our experience enrichments at pH 10–11 yield most alkalophiles, and enrichments which have been kept in excess of pH 11·0 seldom yield organisms.

Enrichment of the following three groups is outlined as an example of the system. Our experience with highly alkaline environments is restricted to African soda lakes which yield a variety of types of bacteria including alkalophilic aerobic heterotrophs, alkalophilic photosynthetic bacteria, and alkalophilic halophiles. We have applied enrichments for alkalophilic heterotrophic bacteria to a variety of ordinary soil types with success.

Heterotrophic bacteria

Several different types of semi-defined media are possible, including nutrient broth, tryptic soy broth (Souza & Deal 1977) adjusted to the appropriate pH value. The most useful general medium is a modification of that described by Horikoshi (1971a).

Composition (gl^{-1}): glucose, 10·0; peptone (Difco), 5·0; yeast extract (Difco), 5·0; K_2HPO_4, 1·0; $MgSO_4.7H_2O$, 0·2; distilled water to 900 ml. The medium is sterilized by autoclaving. Agar at 1·5 % (w/v) may be included if desired. 100 ml 20 % (w/v) $Na_2CO_3.10H_2O$ (sterilized separately by autoclaving) is added to give a final concentration of $Na_2CO_3.10H_2O$ in the medium of 2 % and a final pH of 10·5. Agar containing media should be mixed with $Na_2CO_3.10H_2O$ at 60°C and when mixed poured immediately. A variety of modifications of this medium have been used by Horikoshi and his coworkers to isolate bacteria capable of degrading different carbon sources (Horikoshi 1972; Nakamura & Horikoshi 1976; Ikura & Horikoshi 1977).

A small amount of inoculum is added to the media and incubation is carried out at 25°C or 37°C with shaking for 24–48 hours. Material is then streaked on agar media of the same composition. Both alkaline muds and ordinary soils yield a variety of alkalophilic aerobic heterotrophs by this procedure.

Photosynthetic bacteria

A modification of the medium described by Grant et al. (1979) is suitable for the enrichment and purification of several morphological types of photosynthetic bacteria. As yet these have only been isolated from soda lakes, and enrichments using ordinary river and pond muds have not yielded organisms at high pH.

Composition (gl^{-1}): NH$_4$Cl, 1·0; NaCl, 0·4; MgSO$_4$.7H$_2$O, 0·4; CaCl$_2$.2H$_2$O, 0·5; KH$_2$PO$_4$, 1·0; sodium acetate, 1·0; yeast extract (Difco) 1·0. Distilled water to 900 ml. A solution of trace elements containing (gl^{-1}): ZnSO$_4$.7H$_2$O, 5·5; FeSO$_4$.7H$_2$O, 5·0; Na$_2$MoO$_4$.2H$_2$O, 1·0; MnCl$_2$.4H$_2$O, 5·5; CuSO$_4$, 1·5; CoCl$_2$, 1·5 is added to the medium at 1 ml l^{-1}.

The medium is sterilized by autoclaving. Agar at 1·5% (w/v) may be included. 100 ml 20% (w/v) Na$_2$CO$_3$.10H$_2$O (sterilized separately by autoclaving) is added to give a final concentration of Na$_2$CO$_3$.10H$_2$O of 2% (w/v) and pH 10·3. Na$_2$S at 0·1% may be included in the medium (sterilized separately by autoclaving) if desired, and the NaCl content may be increased depending on the salinity of the environment. Agar containing media should be mixed with Na$_2$CO$_3$.10H$_2$O at 60°C and poured immediately.

Imhoff & Trüper (1977) have used a high salt medium for the isolation of an alkalophilic halophilic *Ectothiorhodospira* sp. from an Egyptian soda lake.

This medium contains (gl^{-1}): CaCl$_2$.2H$_2$O, 0·05; KH$_2$PO$_4$, 0·5; NH$_4$Cl, 0·8; MgCl$_2$.6H$_2$O, 0·1; NaCl, 180·0; Na$_2$SO$_4$, 20·0; Na$_2$CO$_3$, 20·0; yeast extract, 0·5; sodium succinate, 1·0; Na$_2$S, 1·0.

A solution of trace elements at 1 ml l^{-1} of the following composition (gl^{-1}): FeCl$_2$.4H$_2$O, 1·8; CoCl$_2$.6H$_2$O, 0·25; NiCl$_2$, 0·01; CuCl$_2$, 0·01; MnCl$_2$.4H$_2$O, 0·07; ZnCl$_2$, 0·10; H$_3$BO$_3$, 0·5; Na$_2$MoO$_4$.2H$_2$O, 0·03; Na$_2$SeO$_3$.5H$_2$O, 0·01 is added to the medium, followed by 1 ml l^{-1} of vitamin solution of the following composition (gl^{-1}): biotin, 0·1; nicotinamide, 0·35; thiamine, 0·3; *p*-aminobenzoic acid, 0·2; pyridoxal chloride, 0·1; Ca-pantothenate, 0·1; Vit B$_{12}$, 0·05.

The final pH of the enrichment is 9·6.

The initial inoculum is 25% (v/v) mud and incubation is at 30°C at 5000 lx in bottles completely filled. Subculture may be carried out in media of the same composition, incorporating 0·05% (w/v) ascorbate if organic acids alone are to be used as hydrogen donors. Purification can be achieved by repeated agar shake dilution, or alternatively by streaking on agar media in 25 mm plastic Petri dishes, and incubating these in transparent polycarbonate jars (GasPak Anaerobic Systems: Becton, Dickinson and Company, Cockerysville, MD 21030, USA) in the anaerobic environment produced by the GasPak system. The Petri dishes may be held upright round the walls of the jar by an inner glass sleeve, thus affording much better illumination of the plates (Fig. 1). The transparent anaerobic glove box system described by Aranki *et al.* (1969) with an atmosphere of 85% O_2-free N_2/5% CO_2/10% H_2 is particularly useful if large numbers of tests are to be made, and liquid media in divided dishes can be used in this system. Engelhard D catalyst is used in such a system to maintain an O_2 free environment (Engelhard Ltd., Valley Road, Cinderford, Gloucestershire, GL14 2PB) (Fig. 2).

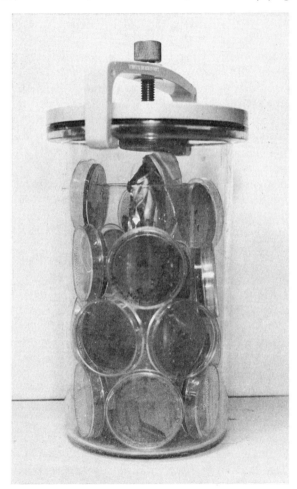

Fig. 1. GasPak polycarbonate jar with glass liner.

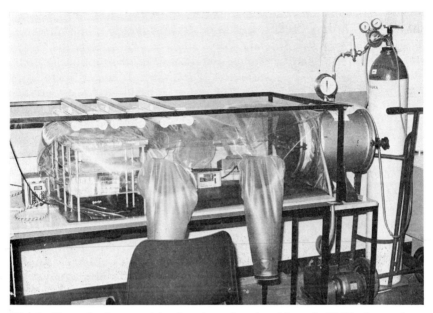

Fig. 2. Clear plastic anaerobic glove box after Aranki *et al.* (1969). Atmosphere 85% O_2 free N_2/5% CO_2/10% H_2. Catalyst Engelhard D. Access by flushable port.

Halophilic bacteria

Extreme examples of soda lakes such as the Wadi Natrun and Lake Magadi yield alkalophiles which are additionally extreme halophiles. One of these is the photosynthetic *Ectothiorhodospira* sp. described by Imhoff & Trüper (1977), the enrichment medium for which has already been described. The solar evaporation ponds at Lake Magadi which contain 16% (w/v) Na_2CO_3, and 16% (w/v) NaCl yield a red alkalophilic halophile which does not contain bacteriochlorophyll. This organism which is probably related to the genus *Halobacterium* (B. J. Tindall and W. D. Grant, unpublished results) can be enriched for in a medium of the following composition (gl^{-1}): $Na_2CO_3.10H_2O$, 200; NaCl, 200; yeast extract (Difco), 1·0; KCl, 1·0; $MgCl_2$, 1·0; NH_4Cl, 1·0. A solution containing 200 g NaCl and 200 g $Na_2CO_3.10H_2O$ is sterilized by autoclaving in a total volume of 500 ml. The other components (including 15 g agar if desired) are also sterilized in a total volume of 500 ml, and the two solutions mixed. Agar containing media should be mixed at 60°C and poured immediately. The pH reading obtained for the medium using sodium corrected probes is 10·6–10·8.

The organism can also be grown in a standard halophile medium adjusted to the appropriate pH value (Brown 1963).

Composition: (gl^{-1}): MgSO$_4$, 20·0; sodium citrate, 3·0; KCl, 2·0; peptone (Difco), 1·0. Make up to 500 ml with distilled water. Sterilization is by autoclaving. A solution containing 250 g NaCl and 50 g Na$_2$CO$_3$.10H$_2$O in 500 ml distilled water is sterilized by autoclaving, and added to the other salts. The final pH of the medium is 10·0. Agar may be incorporated at 1·5 % (w/v) and sterilized with the latter solution.

Enrichment is in shallow static layers of medium at 37°C for 3–4 weeks at 2000–5000 lx. Purification is carried out by streaking on agar media in Petri dishes which are incubated in the light at 37°C. It is necessary to seal the edges of the Petri dishes with transparent adhesive tape to prevent drying out and crystallization of the salt.

Determination of pH Optima

The use of pH gradient plates

The pH range of growth of certain isolates may be rapidly determined using pH gradient plates. Square Petri dishes are suitable for the purpose, and media of the appropriate composition but without Na$_2$CO$_3$.10H$_2$O is poured to a depth of at least 1 cm and allowed to set. A uniform trough 1 cm in width is removed from one end and agar containing 20% Na$_2$CO$_3$.10H$_2$O and 0·2 M NaOH is poured into the trough. The trough agar is prepared by mixing together equal volumes of sterile 0·4 M NaOH/40% (w/v) Na$_2$CO$_3$.10H$_2$O and 4% (w/v) agar at 60°C. Plates left at 37°C overnight develop a uniform gradient from pH 12–pH 7·0 which is stable for 2 days. Organisms are streaked across the plate at right angles to the trough. After 48 h the pH at the alkaline end drops due to CO$_2$ absorption, and therefore if prolonged incubation is necessary the technique is not as useful for determining the upper limits of pH, although useful information on the lower limits can still be obtained. Plates incubated in a CO$_2$ charged atmosphere, e.g. GasPak system, show a more rapid equilibration of the gradient, and the technique only yields information about the lower limits of growth. The surface pH of such plates can be determined using a suitable membrane electrode such as the EIL 1070-2 flat head electrode. The technique is particularly useful for fast growing heterotrophs, and often the pH optima can be predicted from the appearance of the growth. Figure 3 shows isolates from an ordinary soil and an alkaline mud, and clearly shows that one strain is alkalotolerant, whereas the others are alkalophilic.

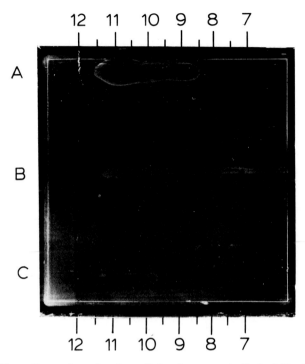

Fig. 3. pH gradient plate showing alkalotolerant and alkalophilic heterotrophs. Strain B is an alkalotolerant isolate from a British garden soil, strain A an alkalophilic isolate from an alkaline mud sample, and strain C an alkalophilic isolate from a British garden soil. The medium is the modified Horikoshi medium (1971a) described in the text.

Buffer systems for high pH

The naturally occurring buffer system in most alkaline environments is the $NaHCO_3/Na_2CO_3$ system. However the system is unsatisfactory for use in determination of pH optima due to the rapid absorption of CO_2 from the atmosphere, which tends to bring high pH media down to between pH 9 and 10 even if relatively high concentrations of Na_2CO_3 have been used. It may be no accident that most of the alkalophiles isolated from naturally occurring environments have pH optima between 9 and 10. The alkaline soda lakes presumably maintain pH values above 10 by constant evaporative concentration and generation of alkalinity by photosynthesis. The most useful buffers over the range 9–12 in a variety of media are borax/NaOH and $Na_2HPO_4/NaOH$. These systems are less effected by atmospheric CO_2 and are non-toxic to a wide range of alkalophiles. 0·2 M solutions of the buffers at appropriate pH's are sterilized separately. Media are brought to the same pH with NaOH and buffer added to a final concentration of 0·05 M. For pH values below 9, NaH_2PO_4/Na_2HPO_4 and borax/boric acid may be used.

References

ALLEN, E. K. & ALLEN, O. N. 1950 Biochemical and symbiotic properties of the rhizobia. *Bacteriological Reviews* **14**, 273–330.

ARANKI, A., SYED, S. A., KENNEY, E. B. & FRETER, R. 1969 Isolation of anaerobic bacteria from human gingiva and mouse cecum by means of a simplified glove box procedure. *Applied Microbiology* **17**, 568–576.

BORNSIDE, G. H. & KALLIO, R. E. 1956 Urea-hydrolysing bacteria I. A physiological approach to identification. *Journal of Bacteriology* **71**, 627–634.

BOYER, E. W. & INGLE, M. B. 1972 Extracellular alkaline amylase from a *Bacillus* species. *Journal of Bacteriology* **110**, 992–1000.

BROWN, A. D. 1963 The peripheral structures of gram-negative bacteria IV. The cation-sensitive dissolution of the cell membrane of *Halobacterium halobium*. *Biochimica et Biophysica Acta* **75**, 425–435.

CHESBRO, W. R. & EVANS, J. B. 1959 Factors affecting the growth of enterococci in highly alkaline media. *Journal of Bacteriology* **78**, 858–862.

CHISLETT, M. E. & KUSHNER, D. J. 1961 A strain of *Bacillus circulans* capable of growing under highly alkaline conditions. *Journal of General Microbiology* **24**, 187–190.

DEAL, P. H., SOUZA, K. A. & MACK, H. M. 1975 High pH, NH_3 toxicity and the search for life on Jovian planets. *Origins of Life* **6**, 561–573.

FOGG, G. E. 1956 The comparative physiology and biochemistry of the bluegreen algae. *Bacteriological Reviews* **20**, 148–165.

GOLTERMAN, H. L. & CLYMO, R. S. 1971 *Methods for Chemical Analysis of Freshwater*. IBP Handbook No. 8, p. 142, Oxford and Edinburgh: Blackwell.

GRANT, W. D., MILLS, A. A. & SCHOFIELD, A. K. 1979. An alkalophilic species of *Ectothiorhodospira* from a Kenyan soda lake. *Journal of General Microbiology* **110**, 137–142.

GUFFANTI, A. A., SUSMAN, P., BLANCO, R. & KRULWICH, T. A. 1978 The protonmotive force and α-aminoisobutyric acid transport in an obligately alkalophilic bacterium. *Journal of Biological Chemistry* **253**, 708–715.

HARDIE, L. A. & EUGSTER, H. P. 1970 The evolution of closed basin brines. *Minerological Society of America, Special Publication* **3**, 273–290.

HORIKOSHI, K. 1971*a* Production of alkaline enzymes by alkalophilic microorganisms. Part I, Alkaline protease produced by *Bacillus*, No. 221. *Agricultural and Biological Chemistry* (*Tokyo*) **35**, 1407–1414.

HORIKOSHI, K. 1971*b* Production of alkaline enzymes by alkalophilic microorganisms. Part II, Alkaline amylase produced by *Bacillus* No. A-40-Z. *Agricultural and Biological Chemistry* (*Tokyo*) **35**, 1783–1791.

HORIKOSHI, K. 1972 Alkaline pectinase of *Bacillus*, No. P-4-N. *Agricultural and Biological Chemistry* (*Tokyo*) **36**, 285–293.

IKURA, Y. & HORIKOSHI, K. 1977 Isolation and some properties of alkalophilic bacteria utilising rayon waste. *Agricultural and Biological Chemistry* (*Tokyo*) **41**, 1373–1377.

IMHOFF, J. F. & TRÜPER, H. G. 1977 *Ectothiorhodospira halochloris* sp. nov. a new extremely halophilic phototrophic bacterium containing bacteriochlorophyll *b*. *Archives of Microbiology* **114**, 115–121.

ISACHENKO, B. L. 1951 *Selected Works* Volume II, pp. 143–162, Moscow: Academy of Sciences of the Union of Soviet Socialist Republics.

KINGSBURY, J. M. 1954 On the isolation, physiology and development of a minute blue green algae. *Ph.D. Thesis*, Harvard University.

KOYAMA, N., KIYOMIYA, A. & NOSOH, Y. 1976 Na^+-dependent uptake of amino acids by an alkalophilic bacillus. *FEBS Letters* **72**, 77–78.

KUSHNER, D. J. 1971 Life in extreme environments. In *Chemical Evolution and the Origin of Life* eds R. Buvet and C. Ponnamperuma. pp. 485–491, London: North Holland.

LANGWORTHY, T. A. 1978 Life in extreme pH values In *Microbial Life in Extreme Environments* ed. D. J. Kushner pp. 279–315, New York: Academic Press.

MELACK, J. M. & KILHAM, P. 1974 Photosynthetic rates of phytoplankton in East African saline lakes. *Limnology and Oceanography* **19**, 743–755.

NAKAMURA, N. & HORIKOSHI, K. 1976 Characterisation and some cultural conditions of a cyclodextrin glycosyltransferase-producing alkalophilic *Bacillus* sp. *Agricultural and Biological Chemistry (Tokyo)* **40**, 753–757.

OHTA, K., KIYOMIYA, A., KOYAMA, N., & NOSOH, Y. 1975 The basis of the alkalophilic property of a species of *Bacillus*. *Journal of General Microbiology* **86**, 259–266.

RUPELA, O. P. & TAURO, P. 1973 Isolation and characterisation of *Thiobacillus* from alkali soils. *Soil Biology and Biochemistry* **5**, 891–897.

SIEGEL, B. Z. 1977 *Kakebekia*. A review of its physiology and environmental features and their relation to its possible ancient affinities. In *Chemical Evolution of the Early Precambrian* ed. C. Ponnamperuma, pp. 143–154, New York: Academic Press.

SIEGEL, S. M. & GIUMARRO, C. 1965 Survival and growth of terrestrial micro-organisms in ammonia high atmospheres. *Icarus* **4**, 37–40.

SIEGEL, B. Z. & SIEGEL, S. M. 1970 Biology of the precambrian genus *Kakabekia* New observations on living *Kakabekia barghoorniana*. *Proceedings of the National Academy of Sciences* **67**, 1005–1010.

SKINNER, F. A. 1968 Limits of microbial existence. *Proceedings of the Royal Society Series B. Biological Sciences* **171**, 77–89.

SOUZA, K. A. & DEAL, P. H. 1977 Characterisation of a novel extremely alkalophilic bacterium. *Journal of General Microbiology* **101**, 103–105.

SOUZA, K. A., DEAL, P. H., MACK, H. M. & TURNBILL, C. E. 1974 Growth and reproduction of micro-organisms under extremely alkaline conditions. *Applied Microbiology* **28**, 1066–1068.

TALLING, J. F. & TALLING, I. B. 1965 The chemical composition of African lake waters. *Internationale Revue der Gesampten Hydrobiologie und Hydrographie* **50**, 421–463.

TE-PANG, H. 1890 *Manufacture of Soda, with Special Reference to the Ammonia Process: a Practical Treatise*. American Chemical Society Monograph No. 65. New York: Hafner Publishing Company.

VEDDER, A. 1934 *Bacillus alcalophilus* n.sp benevens enkele ervaringen metsterk alkalische voedingsbodems *Antonie van Leeuwenhoek* **1**, 141–147.

VALLENTYNE, J. R. 1963 Environmental biophysics and microbial ubiquity. *Annals of the New York Academy of Sciences* **108,** 342–352.

WILEY, W. R. & STOKES, J. C. 1962 Requirement of an alkaline pH and ammonia for substrate oxidation by *Bacillus pasteurii. Journal of Bacteriology* **84,** 730–734.

YAMAMOTO, M., TANAKA, Y. & HORIKOSHI, K. 1972 Alkaline amylases of alkalophilic bacteria. *Agricultural and Biological Chemistry (Tokyo)* **36,** 1819–1823.

Metal-tolerant Micro-organisms of Hot, Acid Environments

CORALE L. BRIERLEY AND J. A. BRIERLEY

*New Mexico Institute of Mining and Technology,
Socorro, New Mexico, USA*

P. R. NORRIS AND D. P. KELLY

*Department of Environmental Sciences,
University of Warwick, Coventry, UK*

High temperature, extremes of acidity or alkalinity and the presence of toxic metal ions are separately extremely restrictive of the diversity of micro-organisms in environments exhibiting such properties. In combination, community diversity is even more severely restricted by these conditions, with relatively few organisms known from hot, acid environments. At least some of those that have been isolated are remarkable also in exhibiting tolerance to high concentrations of metals and in being chemolithotrophic and in some cases also autotrophic.

The acidophilic thermophiles with which we are concerned in this chapter have all been first described since 1966 and are as yet of somewhat uncertain taxonomic position. For convenience we here treat them as two groups: the extreme thermophiles of the *Sulfolobus*-type (Brock *et al.* 1972), having unusual morphology and growing at temperatures up to 85°C; and the moderate thermophiles, generally rod-shaped eubacteria, growing at temperatures up to 55°C (Brierley *et al.* 1978). The breadth of distribution and natural environmental significance of these groups is as yet uncertain, but we shall summarize the present state of knowledge on the properties and culture of these organisms.

Habitats

The acidophilic, thermophilic *Sulfolobus* bacteria have been found in acid thermal springs at temperatures above 60°C in Yellowstone National

Park, USA (Brock *et al.* 1972; Brierley & Brierley 1973) and also in hot, sulphur-rich, acidic soils of this area (Fliermans & Brock 1972). *Sulfolobus*-like organisms have been isolated from an Italian thermal spring area near Naples (De Rosa *et al.* 1974) and from New Zealand hot springs (Bohlool 1975). *Sulfolobus* reportedly occurs in environments with a pH between 2 and 3 and temperatures above 55°C where an oxidizable energy source of either sulphur or ferrous iron is present. These organisms have not yet been reported to exist in metal leaching environments; but another acidophilic extreme thermophile, *Thermoplasma acidophila*, has been isolated from a coal refuse pile (Darland *et al.* 1970).

Acidophilic, moderately thermophilic, *Thiobacillus*-like bacteria have been found in thermal areas of Iceland (Le Roux *et al.* 1977) and Yellowstone National Park USA (Brock *et al.* 1976). This type of microbe has also been found in a test copper-leaching system in which the temperature increased to near 60°C (Brierley & Lockwood 1977), in a copper leach dump in southwestern New Mexico (J. A. Brierley 1978), in ore deposits in the USSR (Golovacheva & Karavaiko 1977), in Bulgaria (Groudev *et al.* 1978), and in samples from the Bingham Mine leach dump, Utah. The samples of the mine dumps were collected from exposed surfaces where the temperature would not be expected to deviate greatly from ambient air temperature. However, these moderately thermophilic strains may be metabolically active in the 'cooler' zones of a leach dump environment, as they have a rather broad temperature range, being able to grow at temperatures at least as low as 30°C and up to a temperature of about 55°C (Brierley & Le Roux 1977). These organisms could be widely distributed in environments of about 50°C and pH 2 where either ferrous iron or a metal sulphide was available as an energy source.

Physiology

Moderate thermophiles

Only one strain, designated TH1, obtained from an Icelandic hot spring (Le Roux *et al.* 1977), has so far been the subject of extensive research dealing with this group's physiology (Brierley & Le Roux 1977; Brierley *et al.* 1978). This organism has some physiological similarities to the mesophilic *Thiobacillus ferrooxidans* in that it oxidizes ferrous iron, pyrite (FeS_2) and metal sulphides at pH 1·4–3·0 but, unlike *T. ferrooxidans*, at temperatures up to 55°C. The TH1 strain can oxidize chalcopyrite and covellite (in the presence of soluble iron) with the concomitant solubilization of copper. Growth on any of these compounds by strains TH1, TH2 (Brierley & Lockwood 1977) or TH3 (J. A. Brierley 1978) in contrast to *T. ferrooxidans*, requires reduced sulphur (e.g. thiosulphate, sulphur or sulphides, such as pyrite) and a source of organic carbon, such as glucose. Yeast extract, cysteine or glutathione can fulfil both the sulphur and carbon requirement, allowing growth and simultaneous oxidation of ferrous iron without the occurrence of significant carbon dioxide fixation (Table 1). Under these conditions its growth is thus mixotrophic, energy being obtained from iron oxidation and carbon from organic sources. These organisms are incapable of autotrophic growth on ferrous sulphate or pyrite in the absence of organic supplements and by comparison with truly autotrophic *T. ferrooxidans* or *Leptospirillum ferrooxidans* (Balashova *et al.* 1974) are virtually incapable of using carbon dioxide as a source of cellular carbon (Table 1). Similarly, while there is no doubt that these strains oxidize ferrous iron, supplied either as soluble ion or as pyrite, the apparent inability to grow consistently on sulphur and the inability of at least some strains to grow on copper sulphides in the absence of oxidizable iron, may all indicate that they can obtain energy only from iron oxidation

TABLE 1

The non-autotrophic nature of thermophile strain TH1

Organism and supplement to medium	Growth temperature ($^{\circ}$C)	Growth yield (mg protein mol^{-1} FeSO$_4$ oxidized)	$^{14}CO_2$ fixed* (c.p.m. mg^{-1} protein mol^{-1} Fe SO$_4$ oxidized)	Proportion of cell-C derived from CO_2 (%)
TH1				
+no supplement	50	0	0	0
+0·02% yeast extract	50	180·7	0·3	1
+0·002% glutathione	50	33·6	1·7	4
+0·02% yeast extract	30	147·8	0·8	2
+0·002% glutathione	30	22·4	0·8	2
T. ferrooxidans				
+no supplement	30	60·3	37·9	100
+0·02% yeast extract	30	119·2	15·4	41
+0·002% glutathione	30	57·4	38·4	100
L. ferrooxidans				
+no supplement	30	45·8	37·2	100
+0·02% yeast extract	30	59·3	27·1	72

*$^{14}CO_2$ specific activity, 382 700 c.p.m. μmol^{-1}

Although it is chemolithotrophic, using iron or pyrite oxidation for energy, this organism fixes only small amounts of carbon dioxide and is obligatorily dependent on organic carbon and reduced sulphur for growth. This can be illustrated by growing the organism in the presence of ^{14}C-carbon dioxide on a medium containing 0·1 M FeSO$_4$ at pH 1·5 and comparing growth and carbon dioxide fixation with *Thiobacillus ferrooxidans* and *Leptospirillum ferrooxidans* in the same medium. (Data of M. Eccleston.)

and not from the oxidation of sulphur or sulphide ion. The well-known chemical reaction of ferric iron with metal sulphides (Kelly *et al.* 1979) to produce ferrous iron could then be the basis of growth on metal sulphides.

In addition to being able to grow as mixotrophs, using chemolithotrophic energy-generating mechanisms and heterotrophic carbon metabolism, the TH strains so far studied (TH1 and TH3) are capable of chemo-organotrophic growth on yeast extract in the absence of metal sulphides or ferrous iron.

Extreme thermophiles

The extremely thermophilic bacteria of the *Sulfolobus*-type are chemolithotrophs, obtaining energy from the oxidation of sulphur, ferrous iron or metal sulphides (Brock *et al.* 1972; Brierley & Brierley 1973; Brierley 1977). At least some isolates are autotrophic, obtaining all their carbon from carbon dioxide fixation (Brock *et al.* 1972). Most strains are capable of heterotrophic growth on yeast extract and other organic substrates (Brock *et al.* 1972; De Rosa *et al.* 1975; C. L. Brierley 1978). Some strains showed enhancement of growth on sulphur or iron in the presence of yeast extract (Brierley & Brierley 1973), while others showed decreased sulphur oxidation although they gave abundant growth (De Rosa *et al.* 1975). Such results indicate varying degrees of mixotrophic capacity: those strains showing decreased inorganic substrate oxidation in the presence of organic supplements being the ones capable of ready heterotrophic growth and presumably capable of greater repression of the chemolithotrophic and autotrophic metabolic mechanisms. The optimum temperature for growth of all these strains is around 70–75°C at pH 2–3 although some strains are able to oxidize sulphur at 84° and probably at 90°C and pH 1·0 (Shivvers & Brock 1973). Apart from their unusual physiological adaptation, these organisms are remarkable for their morphology: they are small, spherical cells, with a primitive sub-microscopic morphology, lacking true cell walls and being surrounded by a plasma membrane and a very fine extracellular coat; some strains show intracellular bodies of an unidentified nature (Brock *et al.* 1972; Brierley & Brierley 1973; Millonig *et al.* 1975). Pili are apparently involved in the attachment of the bacteria to substrates such as sulphur (Weiss 1973).

Taxonomy

"*Sulfolobus*-like" bacteria with the above physiological and morphological characteristics have been described from a number of sources, but analysis of DNA base composition has revealed a range from 39–60 mol % G + C in different isolates (Brock *et al.* 1972; Brierley & Brierley 1973; De Rosa *et al.* 1974, 1975). This indicates that while these organisms share similar morphology, physiology and habitats, they are genetically far from close relatives and almost certainly represent lines of convergent evolution in the colonization of their extreme environments. In view of this, the proposal of De Rosa *et al.* (1974, 1975) that all these types should be lumped in a 'form/habitat' group called *Caldariella* is probably a good one, pending further biochemical study of the strains.

Similarly, the moderately thermophilic iron-oxidizing bacteria cannot easily be classified into any existing genus. There are size and small physiological differences between the strains at present known, although they seem likely to be sufficiently similar to be placed in one genus.

Base composition of DNA has so far been determined only for TH1 as 47–48 mol % G +C. Poor (or lack of) oxidation and growth on sulphur and its compounds in the absence of iron, and failure to grow as chemolithotrophic autotrophs, would make their classification as thiobacilli undesirable. If more rigorous study proves that a wide range of similar isolates are all facultatively heterotrophic mixotrophs, capable of obtaining energy from iron but not sulphur oxidation, then their classification into a new genus would be necessary.

Metal Leaching and Tolerance

The occurrence of the moderately thermophilic eubacteria in leaching environments and their ability to degrade minerals indicate they might have a role in the biogenic leaching of metal sulphides. They exhibit tolerance to metals (e.g. copper, iron, cadmium) similar to that of *T. ferrooxidans* and are similarly inhibited by low concentrations of uranyl or molybdate ion (Tuovinen *et al.* 1971; Brierley *et al.* 1978).

Sulfolobus can leach copper from chalcocite (Brierley 1977) and molybdenum from molybdenite (Brierley 1974) and can grow when the concentration of molybdenum is 7·82 mM, which far exceeds the molybdenum tolerance of *T. ferrooxidans*, which is between 0·05 and 0·9 mM (Tuovinen *et al.* 1971). In one study (Brierley & Brierley 1978), *Sulfolobus* was more effective in leaching chalcopyrite than was *T. ferrooxidans*.

The metal resistance exhibited by these acidophilic organisms almost certainly results mainly from reduced interaction of the organisms with the metals, because of competition for metal-sensitive sites between toxic metal cations and the excess H + ions usually present in the habitat, rather than from any inherent metal-resistant features of the metabolism. A similar phenomenon is seen with other organisms capable of growth at low pH, such as the acid-tolerant fungus *Scytalidium*, which grows in media below pH 2 saturated with copper sulphate, but is sensitive to 2·5 mg Cu per litre at pH 6·7 (Starkey 1973).

Field Sampling Methods

Suitable areas for sampling for acidophilic, thermophilic bacteria would be aerobic, have pH of about 2 to 3 and a temperature of ambient to about 60°C for the moderate thermophiles and 55 to 80°C for the *Sulfolobus* spp.

Leach dump environments which are acidic, have oxidizable sulphides, and possible subsurface temperatures as high as 60 to 80°C (Beck 1967; J. A. Brierley 1978) may have moderate thermophiles present. Samples for study should be taken from the dump *surface* in actively leached areas and should consist of leaching solution (if available) plus mineral/gangue fines. The metal-containing effluent leach solution from the dump should also be sampled for the acidophilic, thermophilic bacteria. *Thiobacillus ferrooxidans* has been shown to be present at depth within leach dumps (Bhappu *et al.* 1969) so core samples, particularly from areas of elevated temperatures, may also contain the moderate thermophiles.

Thermal springs of acid pH are another potential source of *Sulfolobus* and the moderate thermophiles. Samples containing spring water and the sediments of the spring basin or effluent channel should be collected from areas of different temperature. Many springs have effluents with an established temperature gradient. The cooler zones of less than 60°C may have the moderate thermophiles while zones above 60°C may have the *Sulfolobus*-like bacteria.

Techniques for Laboratory Culture of Thermophiles

Commonly, cultures are grown in shaken Erlenmeyer flasks, containing 30–50% by volume of medium, which allows adequate aeration. When used as an energy source, the solid mineral concentration seldom exceeds 10% (w/v). Flasks for thermophilic studies can be capped with aluminium foil to reduce evaporation. Since incubation is at 50°C or higher, evaporation can be considerable in a dry air incubator, and incubation in water baths is recommended.

Routine cultivation of TH strains in 100 ml medium volumes in 250 ml flasks shaken in an orbital dry air incubator does not lead to excessive evaporation over the short periods generally required for growth on ferrous iron with yeast extract (1–3 days) and correction for evaporation losses over longer periods can be made by periodically weighing the flasks and replacing evaporate with sterile distilled water. If cultures are aerated by sparging, care must be taken to saturate the air with water to avoid excessive evaporation. In cultivation in fermenters or during continuous chemostat cultivation of thermophiles a cold-water condenser should be fitted to return evaporated water to the culture vessel. The addition of carbon dioxide up to 2% (v/v) of the gas phase has been used to enhance growth of mesophilic thiobacilli (Bryner *et al*. 1954; Schnaitman & Lundgren 1965) and this technique could readily be applied to the thermophilic organisms, *Sulfolobus* in particular, which has been shown to give faster growth and sulphur oxidation under increased partial pressures of CO_2 (Shivvers & Brock 1973).

Media

The 9 K medium of Silverman & Lundgren (1959) has been extensively used for studies of iron oxidation by *T. ferrooxidans*. However, this medium at pH 2·8 is unsatisfactory for use in studies of thermophilic, iron-oxidizing bacteria as the ferrous iron spontaneously oxidizes rapidly at temperatures above 50°C, with concomitant formation of large amounts of basic ferric sulphates, ferric hydroxide and jarosite. Ferric iron precipitates entrap the bacteria reducing the number of organisms in solution. Adjusted to about pH 1·5, this medium should be suitable for thermophiles, but we have had quite satisfactory results using modified media containing less iron and ammonium sulphate. The medium used by Tuovinen & Kelly (1973) for *T. ferrooxidans* is very suitable for the moderate thermophiles when adjusted to pH 1·5–1·6 to prevent or minimize ferric iron precipitation and supplemented with 0·02% (w/v) yeast extract. MacKintosh (1978) has used a "low phosphate" salts medium in order to reduce the amount of precipitated ferric sulphate and the subsequent loss of microbes. A "low phosphate" modification of the Tuovinen & Kelly medium (1973), practical fur culturing of the moderate thermophiles contains (g l^{-1}): K_2HPO_4, 0·40; $MgSO_4.7H_2O$, 0·4; $(NH_4)_2SO_4$, 0·4; $FeSO_4.7H_2O$, 27·8; either yeast extract, 0·2; glutathione, 0·1; or cysteine, 0·1. The medium is adjusted to a pH value between 1·5 and 1·6 using H_2SO_4 and sterilized by autoclaving at either 115°C for 10 minutes or 121°C for 5 minutes. The medium must be supplemented with one of the organic compounds when culturing the moderate thermophile strains. The ferrous sulphate may be omitted, and pyrite, chalcocite or other metallic sulphides may be used for energy sources at a concentration of 10 g l^{-1}. A medium containing pyrite and yeast extract would probably be most suitable for initial isolation of the moderate thermophiles (high ferrous iron concentrations have been observed to inhibit some mesophilic iron-oxidizing organisms until they became adapted to the laboratory growth conditions) and initial experiments have shown that such a medium allows maintenance of healthy stock cultures without the frequent subculturing required if stocks are maintained in ferrous sulphate/yeast extract medium.

The acidophilic, extreme thermophiles of the *Sulfolobus*-type grow well using the medium of Brock *et al.* (1972) or Brierley & Brierley (1973). Although *Sulfolobus* grow well by a strictly chemolithotrophic mode of metabolism, cell yield and apparent rate of growth can be enhanced by supplementing media with 0·02% (w/v) yeast extract.

Optimum temperature reported for growth of a moderately thermophilic strain is near 50°C (Brierley & Le Roux 1977). The very thermophilic, acidophilic organisms, *Sulfolobus acidocaldarius* (Brock *et al.* 1972) and related strains (Brierley & Brierley 1973; De Rosa *et al.* 1975) are reported to have a broad temperature range varying from 55–94°C; however, optimum growth is obtained between 60–70°C.

Culturing of the thermophilic bacteria on agar-hardened media has met with limited success (Brock *et al.* 1972), since none of the bacteria grow well on solid agar, and the heat necessary for development of the thermophiles tends to cause the agar to soften. The membrane filter technique used for culturing *T. ferrooxidans* (Tuovinen & Kelly 1973), might also be applicable to culturing the thermophilic organisms.

Percolator leach columns (Zajic 1969) are used to simulate conditions existing in dump leaching operations. These columns can be used successfully for leaching studies with thermophilic bacteria by heating the columns with heat tapes. Incubation of the entire system in a dry air oven is discouraged because of rapid solution evaporation.

Columns $1·8$ m $\times$ $0·15$ m can be used for leaching evaluation and economic studies (Brierley 1977). The columns and holding tanks are warmed with heat tapes. Leach solution is pumped from the holding tank to the top of the column and is applied to the ore through a circular rose with a series of openings. This ensures even distribution of the solution to the ore surface. The particles used in the leach columns should not be excessively large or channelling of the leach solution will occur.

Enumeration Methods

The most probable number (MPN) technique (Collins 1967; Tuovinen *et al.* 1971; Bhappu *et al.* 1969) can be used for enumeration of thermophilic, chemolithotrophic microbes. There are limitations to this technique since organisms attached to solid mineral particles cannot be removed to obtain a dispersion for quantitative sampling. However, the MPN test will give indicative trends of bacterial populations.

The bacteria are often attached to the solid inorganic substrate, iron

precipitates, and walls of the vessels used for culturing. This causes in-convenience in enumeration of the population by direct methods such as microscopic counting. Consequently, direct enumeration is avoided and more often indirect methods for measuring activity are performed. When sulphur is used as an energy source, a decrease in pH or increase in titratable acidity can be used as a measure of activity. Likewise, the decrease in ferrous iron concentration can be used as a measure of iron-oxidizing activity by the organisms; on mineral substrates the increasing amount of metal solubilized will be indicative of activity. Standard protein measurement methods (Lowry *et al.* 1951) are frequently used to determine bacterial biomass, but this method can be subject to interfer-ence by inorganic ions. The use of turbidity for indication of chemolitho-trophic growth is infrequently used since most growth substrates are themselves particulate or bacterial growth results in the substrates becom-ing particulate (e.g. through the precipitation of iron compounds) or in a colour change in the medium. Bacterial dry weight is not used for the same reason, unless precipitation is prevented by low pH.

Other techniques which have recently been applied to studies of the chemolithotrophic bacteria and have applicability to the thermophilic organisms include oxygen uptake, measured with Clark oxygen electrode cells (Brierley *et al.* 1978) or using a Gilson Differential Respirometer or a Warburg Respirometer according to standard techniques (Umbreit *et al.* 1972); fixation of ^{14}C-labelled CO_2 or organic substrates; total nitrogen estimation; fluorescent antibody staining; ATP assay; and particle counting with a Coulter counter: the latter is subject to error if there is inorganic precipitation in the media.

Smith *et al.* (1972) adapted a technique used in aquatic systems for measuring autotrophic CO_2 fixation to the study of chemolithotrophic bacteria in acid soils. $^{14}CO_2$ is added to soil samples, and time for incorpor-ation of the CO_2 into organisms is allowed. The incubation is stopped, and the ^{14}C-labelled organic matter oxidized. The $^{14}CO_2$ generated is trans-ferred to a phenylamine-liquid scintillation counting system. This technique has been successfully used with mesophilic chemoautotrophs; however, contradictory results have been achieved when the test was applied to *Sulfolobus* in leaching systems (Brierley 1977).

Nitrogen assay can be used as a method for obtaining an estimation of bacterial numbers in an inorganic medium (McGoran *et al.* 1969; Le Roux *et al.* 1973). However, this method does not differentiate between bacterial nitrogen and inorganic nitrogen. A variation of this technique (Gormley & Duncan 1974) prescribes that total nitrogen be determined by Kjeldahl digestion and inorganic nitrogen (distillable nitrogen) be determined by standard alkaline steam-distillation.

The fluorescent antibody (FA) staining technique has been used for detection of *T. ferrooxidans* on coal refuse particles (Apel *et al.* 1976), and this technique has been successfully applied to the study of *Sulfolobus* attachment on glass slides in hot springs (Bohlool & Brock 1974). The major limitation of the technique is the autofluorescence and non-specific fluorescence resulting from application of the FA stain to mineral specimens.

The measurement of adenosine triphosphate (ATP) by the luciferin-luciferase reaction (McElroy 1947) has been applied to measurement of chemolithotrophic biomass (J. A. Brierley, unpublished data).

The Coulter Counter can be used to determine numbers of suspended thiobacilli (Shuler & Tsuchiya 1975); however, it is not applicable to establishing numbers when the organisms are attached to mineral particles. The Coulter Counter has not been applied to estimation of thermophilic microbial populations.

Microscopic Examination

Conventional, simple staining methods used for heterotrophic bacteria can be used for light microscopy.

Transmission electron micrographs (TEM) of thin sections and whole cells of the very thermophilic, acidophilic *Sulfolobus* have been prepared (Brierley & Brierley 1973; Millonig *et al.* 1975). Standard thin section techniques can be applied after a dense suspension of clean cells is obtained. Negative and positive staining for TEM of whole cells can be accomplished, but the dense suspension of clean cells should be suspended in a neutral buffer to avoid precipitation of the staining compounds.

The attachment of *Sulfolobus* to sulphur (Weiss 1973; Baldensperger *et al.* 1974) and molybdenite (Brierley & Murr 1973; Brierley *et al.* 1973) has been studied using scanning electron microscopy. Considerable care must be exercised when examining the attachment of these organisms to inorganic substrates, since artefacts produced by spontaneous oxidation of sulphide minerals can readily be interpreted as attached bacteria and it is sometimes difficult to distinguish cells from inorganic matter (Wyckoff & Davidson 1977). It is essential that sterile controls of mineral specimens be examined to ensure that observed structures are indeed micro-organisms.

References

APEL, W. A., DUGAN, P. R., FILPPI, J. A. & RHEIMS, M. S. 1976 Detection of *Thiobacillus ferrooxidans* in acid mine environments by indirect fluorescent antibody staining. *Applied and Environmental Microbiology* **32**, 159–165.

BALASHOVA, V. V., VEDININA, I. YA., MARKOSYAN, G. E. & ZAVARZIN, G. A. 1974 The auxotrophic growth of *Leptospirillum ferrooxidans*. *Mikrobiologiya* **43**, 581–585 (English translation pp. 491–494.)

BALDENSPERGER, J., GUARRAIA, L. J. & HUMPHREYS, W. J. 1974 Scanning electron microscopy of thiobacilli grown on colloidal sulphur. *Archives of Microbiology* **99**, 323–329.

BECK, J. V. 1967 The role of bacteria in copper mining operations. *Biotechnology and Bioengineering* **9**, 487–497.

BHAPPU, R. B., JOHNSON, P. H., BRIERLEY, J. A. & REYNOLDS, D. H. 1969 Theoretical and practical studies on dump leaching. *Transactions of the Society of Mining Engineers AIME* **244**, 307–320.

BOHLOOL, B. B. 1975 Occurrence of *Sulfolobus acidocaldarius*, an extremely thermophilic acidophilic bacterium, in New Zealand hot springs. Isolation and immunofluorescence characterisation. *Archives of Microbiology* **106**, 171–174.

BOHLOOL, B. B. & BROCK, T. D. 1974 Population ecology of *Sulfolobus acidocaldarius*. II. Immunological studies. *Archives of Microbiology* **97**, 181–194.

BRIERLEY, C. L. 1974 Molybdenite-leaching: use of a high-temperature microbe. *Journal of Less-Common Metals* **36**, 237–247.

BRIERLEY, C. L. 1977 Thermophilic micro-organisms in extraction of metals from ores. *Developments in Industrial Microbiology* **18**, 273–284.

BRIERLEY, C. L. 1978 Bacterial leaching. *CRC Critical Reviews in Microbiology* **5**, 207–262.

BRIERLEY, C. L. & BRIERLEY, J. A. 1973 A chemoautotrophic and thermophilic micro-organism isolated from an acid hot spring. *Canadian Journal of Microbiology* **19**, 183–188.

BRIERLEY, C. L. & MURR, L. E. 1973 Leaching: use of a thermophilic and chemoautotrophic microbe. *Science, New York* **179**, 488–489.

BRIERLEY, C. L., MURR, L. E. & BRIERLEY, J. A. 1973 Using the SEM in mining research. *Research and Development* **24**, 24–28.

BRIERLEY, J. A. 1978 Thermophilic iron-oxidizing bacteria found in copper-leaching dumps. *Applied and Environmental Microbiology* **36**, 523–525.

BRIERLEY, J. A. & BRIERLEY, C. L. 1978 Microbial leaching of copper at ambient and elevated temperatures. In *Metallurgical Applications of Bacterial Leaching and Related Microbiological Phenomena* eds Murr, L. E., Torma, A. E. & Brierley, J. A. pp. 477–490, New York: Academic Press.

BRIERLEY, J. A. & LE ROUX, N. W. 1977 A facultative thermophilic *Thiobacillus*-like bacterium: oxidation of iron and pyrite. *Conference: Bacterial leaching* ed. Schwartz, W. pp. 55–66, Weinheim, New York: Verlag Chemie.

BRIERLEY, J. A. & LOCKWOOD, S. J. 1977 The occurrence of thermophilic iron-oxidizing bacteria in a copper leaching system. *FEMS Microbiology Letters* **2**, 163–165.

BRIERLEY, J. A., NORRIS, P. R., KELLY, D. P. & LE ROUX, N. W. 1978 Characteristics of a moderately thermophilic and acidophilic iron-oxidizing *Thiobacillus*. *European Journal of Applied Microbiology and Biotechnology* **5**, 291–299.

BROCK, T. D., BROCK, K. M., BELLY, R. T. & WEISS, R. L. 1972 *Sulfolobus*:

a new genus of sulfur-oxidizing bacteria living at low pH and high temperature. *Archiv für Mikrobiologie* **84,** 54–68.

BROCK, T. D., COOK, S., PETERSEN, S. & MOSSER, J. L. 1976 Biogeochemistry and bacteriology of ferrous iron oxidation in geothermal habitats. *Geochimica et Cosmochimica Acta* **40,** 493–500.

BRYNER, L. D., BECK, J. V., DAVIS, B. D. & WILSON, D. G. 1954 Micro-organisms in leaching sulfide mineral. *Industrial and Engineering Chemistry* **46,** 2587–2592.

COLLINS, C. H. 1967 *Microbiological Methods* 2nd edn, New York: Plenum Press.

DARLAND, G., BROCK, T. D., SAMSONOFF, W. & CONTI, S. F. 1970 A thermophilic, acidophilic mycoplasma isolated from a coal refuse pile. *Science* **170,** 1416–1418.

DE ROSA, M., GAMBACORTA, A., MILLONIG, G. & BU'LOCK, J. D. 1974 Convergent characters of extremely thermophilic acidophilic bacteria. *Experientia* **30,** 866–868.

DE ROSA, M., GAMBACORTA, A. & BU'LOCK, J. D. 1975 Extremely thermophilic acidophilic bacteria convergent with *Sulfolobus acidocaldarius. Journal of General Microbiology* **86,** 156–164.

FLIERMANS, C. B. & BROCK, T. D. 1972 Ecology of sulfur-oxidizing bacteria in hot acid soils. *Journal of Bacteriology* **111,** 343–350.

GOLOVACHEVA, R. S. & KARAVAIKO, G. I. 1977 A new facultative thermophilic *Thiobacillus* isolated from sulphide ore. In *Microbial Growth on C_1-compounds (abstracts)* pp. 108–109, Puschino: USSR Academy of Sciences.

GORMLEY, L. S. & DUNCAN, D. W. 1974 Estimation of *Thiobacillus ferrooxidans* concentrations. *Canadian Journal of Microbiology* **20,** 1453–1455.

GROUDEV, S. N., GENCHEV, F. N. & GAIDARJIEV, S. S. 1978 Observations on the microflora in an industrial copper leaching operation. In *Metallurgical Applications of Bacterial Leaching and Related Microbiological Phenomena* eds Murr, L. E., Torma, A. E. & Brierley, J. A. pp. 253–274, New York: Academic Press.

KELLY, D. P., NORRIS, P. R. & BRIERLEY, C. L. 1979 Microbiological methods for the extraction and recovery of metals. *Symposium of the Society for General Microbiology* **29,** 263–308.

LE ROUX, N. W., NORTH, A. A. & WILSON, J. C. 1973 Bacterial oxidation of pyrite. *Tenth International Mineral Processing Congress* Paper 45. London: The Institution of Mining and Metallurgy.

LE ROUX, N. W., WAKERLEY, D. S. & HUNT, S. D. 1977 Thermophilic *Thiobacillus*-type bacteria from Icelandic thermal areas. *Journal of General Microbiology* **100,** 197–201.

LOWRY, O. H., ROSEBROUGH, N. J., FARR, A. L. & RANDALL, R. J. 1951 Protein measurement with the Folin phenol reagent. *Journal of Biological Chemistry* **193,** 265–275.

MACKINTOSH, M. E. 1978 Nitrogen fixation by *Thiobacillus ferrooxidans. Journal of General Microbiology* **105,** 215–218.

MCELROY, W. D. 1947 The energy source for bioluminescence in an isolated system. *Proceedings of the National Academy of Sciences, U.S.A.* **33,** 342–345.

MCGORAN, C. J. M., DUNCAN, D. W. & WALDEN, C. C. 1969 Growth of *Thiobacillus ferrooxidans* on various substrates. *Canadian Journal of Microbiology* **15,** 135–138.

MILLONIG, G., DE ROSA, M., GAMBACORTA, A. & BU'LOCK, J. D. 1975 Ultra-structure of an extremely thermophilic acidophilic micro-organism. *Journal of General Microbiology* **86,** 165–173.

SCHNAITMAN, C. & LUNDGREN, D. G. 1965 Organic compounds in the spent medium of *Ferrobacillus ferrooxidans*. *Canadian Journal of Microbiology* **11,** 23–27.

SHIVVERS, D. W. & BROCK, T. D. 1973 Oxidation of elemental sulfur by *Sulfolobus acidocaldarius*. *Journal of Bacteriology* **114,** 706–710.

SHULER, M. L. & TSUCHIYA, H. M. 1975 Determination of cell numbers and cell size for *Thiobacillus ferrooxidans* by the electrical resistance method. *Biotechnology and Bioengineering* **17,** 621–624.

SILVERMAN, M. P. & LUNDGREN, D. G. 1959 Studies on the chemoautotrophic iron bacterium *Ferrobacillus ferrooxidans*. I. An improved medium and a harvesting procedure for securing high cell yields. *Journal of Bacteriology* **77,** 642–647.

SMITH, D. W., FLIERMANS, C. B. & BROCK, T. D. 1972 Technique for measuring $^{14}CO_2$ uptake by soil microorganisms *in situ*. *Applied Microbiology* **23,** 595–600.

STARKEY, R. L. 1973 Effect of pH on toxicity of copper to *Scytalidium* sp., a copper-tolerant fungus, and some other fungi. *Journal of General Microbiology* **78,** 217–225.

TUOVINEN, O. H. & KELLY, D. P. 1973 Studies on the growth of *Thiobacillus ferrooxidans*. *Archiv für Mikrobiologie* **88,** 285–298.

TUOVINEN, O. H., NIEMELA, S. I. & GYLLENBERG, H. G. 1971 Tolerance of *Thiobacillus ferrooxidans* to some metals. *Antonie van Leeuwenhoek Journal of Microbiology and Serology* **37,** 489–496.

UMBREIT, W. W., BURRIS, R. H. & STAUFFER, J. R. 1972 *Manometric and Biochemical Techniques* 5th edn, Minneapolis: Burgess.

WEISS, R. L. 1973 Attachment of bacteria to sulphur in extreme environments. *Journal of General Microbiology* **77,** 501–507.

WYCKOFF, R. W. G. & DAVIDSON, F. D. 1977 The composition, morphology and action upon chalcopyrite of autotrophs recovered from fumaroles. In *Conference: Bacterial Leaching* ed. Schwarz, W. pp. 67–74, Weinheim, New York: Verlag Chemie.

ZAJIC, J. E. 1969 *Microbial Biogeochemistry* New York: Academic Press.

Colonization of Damp Organic Substrates and Spontaneous Heating

Rothamsted Experimental Station, Harpenden, Hertfordshire, UK

A wide range of organic materials can heat spontaneously when stored damp (Table 1). Before storage, they acquire a varied inoculum of bacteria, actinomycetes and fungi by airborne deposition and rainsplash; by contact with soil, with previously colonized matter or contaminated surfaces; by epiphytic growth on the surface of plant material or by passage through the digestive system of animals. Which organisms from this inoculum develop subsequently and the rate of their development depend on the conditions of storage, particularly the water activity of the substrate, aeration, ambient temperature and nutritional factors.

Spontaneous heating is a consequence of the release of energy during the respiration of plant and microbial cells. If this energy cannot escape, as quickly as it is produced during rapid microbial growth, the substrate heats up and conditions within may be modified sufficiently to enable other more thermophilic organisms to grow.

This article describes how different factors affect some of the organisms colonizing damp organic substrates and their spontaneous heating and how they interact to determine the pattern of colonization, using hay as an example.

Methods of Study

The colonization and spontaneous heating of damp organic substrates has been studied at many different levels. Large-scale studies have utilized, for instance, stacks of loose hay (Miehe 1930), baled hay (Gregory *et al.* 1963) or baled sugar-cane bagasse (Lacey 1974), cereal grains stored in elevators or silos (Sinha & Wallace 1965; Lacey 1971), manure piles for mushroom compost (Henssen 1957*a,b*; Fergus 1964), wool (Walker & Williamson 1957; Walker & Harrison 1960), guayule rets (Cooney & Emerson 1964) and woodchip piles (Tansey 1971; Smith & Ofosu-Asiedu 1972). In smaller scale field studies, very small bales of hay

TABLE 1

Substrates supporting spontaneous heating

Agricultural produce

  Hay and straw
  Cereal grains and other seeds
  Fermented foods
  Cocoa, coffee, tea
  Tobacco
  Cotton
  Wool

Forest products

  Wood chip piles

Waste materials

  Sugar-cane bagasse
  Vegetable composts and hot beds
  Manure of herbivorous animals
  Municipal waste

Mushroom compost
Retting guayule
Nests of Mallee fowl

and sacks of cereal grains have been stored in boxes made of polystyrene foam insulation board (Easson & Nash 1978; Westermarck-Rosendahl & Ylimäki 1978). In the laboratory, many studies of different substrates have utilized wide-necked Dewar flasks of 1 to 4·5 litre capacity (Fig. 1) (James 1927; James *et al.* 1928; Miehe 1930; Ramstad & Geddes 1942; Carter & Young 1950; Festenstein *et al.* 1965). However, Carter & Young (1950) point out that such flasks may differ in their insulating

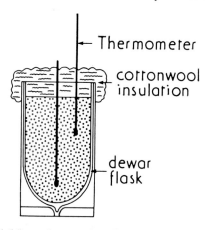

Fig. 1. Dewar flask (4·5 l) used to study colonization and spontaneous heating of hay and grain.

properties and for greater control of conditions, adiabatic incubators maintaining a temperature close to, but slightly below, that of the heating material have been used (Ramstad & Geddes 1942; James & Lejeune 1952; Dye & Rothbaum 1964; Currie & Festenstein 1971). Aeration can be supplied by tube to the base of the container. For critical studies the relative humidity of air entering the flask may be controlled and the water, oxygen and carbon dioxide content of the outflowing air monitored (Dye & Rothbaum 1964; Currie & Festenstein 1971). Alternatively, heat output can be measured at different temperatures under isothermal conditions (Rothbaum & Dye 1964).

Isolation of Micro-organisms from Spontaneously Heated Materials

Traditionally dilution plate methods have been used for the isolation of micro-organisms from spontaneously heated substrates. While these may be satisfactory for isolating and enumerating fungi and bacteria, they may be much less satisfactory for the isolation of actinomycetes that can form a large proportion of the microflora. Thus, an anaerobic technique has been used for isolating actinomycetes from manures (Henssen 1957a; but see p. 57) but aeration is usually necessary. Actinomycetes and fungi with spores that easily become airborne can be isolated satisfactorily by suspending the spores in air either in the airstream of a small wind tunnel or in an enclosed sedimentation chamber. Samples of the air may then be taken using an Andersen sampler to deposit spores on agar media in Petri dishes (Gregory & Lacey 1963; Lacey 1971; Lacey & Dutkiewicz 1976a,b).

Malt extract agar containing streptomycin (40 μg ml^{-1}) and penicillin (20 iu ml^{-1}) have been used satisfactorily for fungi although Littman's oxgall and OAES agars are possible alternatives (Lacey & Dutkiewicz 1976a) while potato dextrose agar and, for thermophiles, YpSs and yeast glucose agars have also been suggested (Dye 1964; Cooney & Emerson 1964). For actinomycetes and bacteria, half-strength nutrient agar or half-strength tryptone soya-casein hydrolysate agar, both containing actidione (50–100 μg ml^{-1}) or tryptone-yeast extract agar may be used (Lacey & Dutkiewicz 1976a; Dye 1964).

Thermophilic organisms are often characteristic of the microflora of spontaneously heated substrates. Consequently a range of incubation temperatures must be used to ensure isolation of all components of the microflora, e.g. 25 and 40°C for fungi and 25, 40 and 55°C for actino-mycetes and bacteria (Lacey & Dutkiewicz 1976a).

Factors Affecting Colonization and Heating

Colonization of heating substrates is affected by their water activity, aeration, temperature, nutritional status and pH and the mass of material and its density.

Water activity

The water activity (a_w) of a substance is defined as the ratio of the vapour pressure of water over a substrate to that of pure water at the same temperature and pressure. Numerically, a_w is the same as equilibrium relative humidity (ERH) but it is expressed as a proportion of one rather than as a percentage. It provides a measure of the availability of water in a substrate to organisms growing on it. At $1 \cdot 0$ a_w, free water is available in the substrate. Water content is, perhaps, a more easily derived figure but for different substances, the same water content may give very different water activities (Snow *et al.* 1944; Ayersts 1965, 1969).

The a_w determines the amount of microbial activity a substrate can support, the species that can grow and consequently the rate of energy release and heating. Most species of fungi require at least $0 \cdot 75$ a_w while no fungal growth occurs with less than $0 \cdot 61$ a_w. *Monascus* (*Xeromyces*) *bisporus* is the most xerophilic fungus known with a minimum a_w for growth of only $0 \cdot 61$ (Pitt & Christian 1968). Above $0 \cdot 65$ a_w on cereal grains species of the *Aspergillus restrictus* and *Aspergillus glaucus* groups and *Wallemia sebi* start to grow with successively *Aspergillus versicolor*, *A. candidus*, *A. flavus* and other species commencing growth as a_w is increased (Christensen & Kaufmann 1974). For most bacteria and actinomycetes an a_w greater than $0 \cdot 93$ is required but some will grow at about $0 \cdot 90$ a_w and a few halophilic species down to $0 \cdot 75$ a_w or less.

Aeration

Maximum energy release from carbohydrates is by aerobic respiration when $C_6H_{12}O_6 + 6O_2 \rightarrow 6H_2O + 6CO_2 + 2800J$. Anaerobic respiration releases little energy and causes little or no heating. Many experiments have shown that heating ceases as soon as oxygen in the intergranular air of sealed containers is exhausted (Miehe 1930; Carter & Young 1950; Festenstein *et al.* 1965). The requirement for oxygen increases with temperature to a maximum at about 40°C, but does not decrease greatly until above 65°C. At this temperature microbial growth is inhibited and most heat results from exothermic chemical oxidation. The respiratory quotient may be $0 \cdot 7$ to $0 \cdot 9$ during the temperature rise to 66°C but is less than $0 \cdot 5$ above this (Miehe 1930; Currie & Festenstein 1971).

Respiration may occur in still-living plant cells during the early stages of heating but is soon surpassed by microbial respiration and ceases at about 40°C when plant cells are killed.

Most fungi and actinomycetes are considered obligate aerobes although some, like yeasts and *Penicillium roqueforti*, can grow in atmospheres deficient in oxygen. Henssen (1957*b*) described some thermophilic fungi and actinomycetes as facultative aerobes on the basis of growth in sealed containers together with alkaline pyrogallol. However, she did not analyse the gases present and Kane & Mullins (1973) and Deploey & Fergus (1975) found subsequently that none grew in pure nitrogen and that more oxygen was necessary for sporulation than for trace growth. Most required at least 0·2% oxygen for trace growth, 0·7 to 1·0% for moderate growth and at least 1·0% for sporulation.

Temperature

Organisms vary widely in their growth temperature ranges and optimum growth temperatures, not only between genera but also between the species of a genus and sometimes between strains of a single species (Fig. 2). Most thrive between 10 and 40°C with optima of 25 to 35°C and

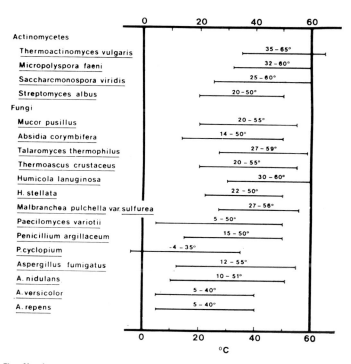

Fig. 2. Cardinal temperatures for growth of some fungi and actinomycetes colonizing damp organic substrates.

are classified as mesophiles. A few grow at lower or higher temperatures. In the context of spontaneous heating, thermophiles able to grow at elevated temperatures are of greatest importance.

The concept of thermophily has been defined differently for different groups of organisms and by different authors. Here, thermophilic fungi are considered as those with a minimum temperature for growth greater than 20°C and a maximum greater than 50°C (Cooney & Emerson 1964). Thermophilic actinomycetes, like other thermophilic bacteria, also have maxima greater than 50°C but their minima may be between 30 and 37°C (Cross 1968). However, in neither group is there a distinct boundary and thermotolerant or thermoduric organisms overlap the mesophilic and thermophilic temperature ranges, e.g. *Aspergillus fumigatus, Absidia corymbifera, Streptomyces albus* (Fig. 2).

In nature, temperature and water activity interact to determine the growth range of individual species. This is illustrated for some fungi in Fig. 3, but there is little or no comparable data for thermophilic fungi or actinomycetes.

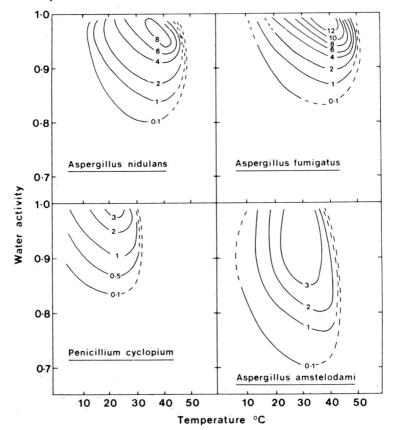

Fig. 3. Growth rate (mm) of four fungi at different temperatures and water activities (after Ayerst 1969).

The part individual organisms can contribute to the heating of the substrate was first demonstrated by Miehe (1930). Growing pure cultures on sterilized breadcrumbs or hay, he demonstrated temperature rises of more than 30°C by some species (Table 2). Similarly, Hussain (1973) showed that hay (water content 50%) inoculated with *Bacillus stearothermophilus* and *Thermoactinomyces vulgaris* would increase in temperature from 30°C to 55–57°C. However, hay inoculated with *B. stearothermophilus* alone continued to heat to 260°C.

Although some species can commence growth close to freezing point, growth is slow and heat output small so that low ambient air temperatures inhibit or delay spontaneous heating (Miehe 1930; Waite 1949). However,

TABLE 2

Spontaneous heating by pure cultures (Miehe 1930)

Species	Minimum temperature for growth °C	Maximum temperature for growth °C	Heated substrate up to °C	
Fungi				
Rhizopus				
nigricans	1	35–40	38	
Aspergillus				
glaucus				
group (?)	?	41	39	
A. niger	7–10	50–55	53·5	
A. fumigatus	15	55–60	57	ambient temperature 40°C
			54	ambient temperature 20°C
Absidia				
corymbifera	20	60	56·5	
Malbranchea				
pulchella var.				
sulfurea	29	55–60	58	
Humicola				
lanuginosa	30	60–65	62·5	
Bacteria				
Bacillus calfactor	30	75–80	51	in hay, ambient temperature 40°C
			68	in hay, ambient temperature 60°C
Thermoactinomyces				
vulgaris (?)				
(as *Actinomyces*				
thermophilus)	30	60–65	63	

heating occurs rapidly from 15–20°C, hay characteristically showing a two-peaked curve reaching a maximum in 3 to 7 days (Fig. 4). The first peak may represent heating from respiring plant cells and mesophilic fungi on the hay at baling and the second the subsequent development of thermoduric and thermophilic organisms. Hay that eventually ignites maintains a temperature near 65–75°C for a period before increasing rapidly to over 200°C (Fig. 5) (Gregory *et al.* 1963; Currie & Festenstein 1971). Microbial heating is limited to a maximum of 65–70°C and unless spontaneous ignition occurs, the maximum temperature of heating is closely related to water content at the start of the experiment (Fig. 6) (Gregory *et al.* 1963; Festenstein *et al.* 1965).

Nutrition

Vigorous microbial growth leading to spontaneous heating depends on the presence of soluble organic substances in the substrate. Soluble

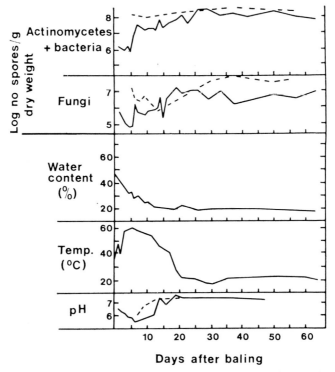

Fig. 4. Colonization, heating and changes in water content and pH in hay baled at 46% water content (after Gregory *et al.* 1963); (- - -), samples from a faster moulding bale.

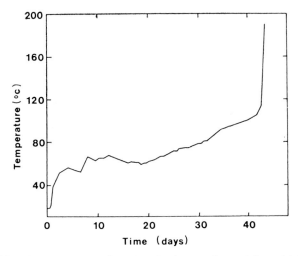

Fig. 5. Self-heating beyond 100°C of 125 g hay moistened in a 1 l Dewar flask incubated adiabatically (after Currie & Festenstein 1971).

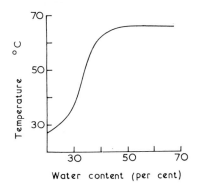

Fig. 6. Maximum temperatures attained by spontaneous heating of hay at different water contents.

carbohydrates form up to 3% of the dry matter of hay (Fig. 7) and more than 5% of that of bagasse. During microbial colonization and spontaneous heating they are rapidly utilized. More complex carbohydrates such as starch, cellulose and lignocellulose may also be utilized by fungi and actinomycetes, including thermophilic species (Gregory *et al.* 1963; Kuo & Hartman 1966; Fergus 1969*a,b*; Hedger & Hudson 1974; Lacey 1974; Oso 1978; Rosenberg 1978).

   Total nitrogen may appear to increase as a result of the loss of dry matter, but the proportion that is soluble decreases steadily during

J. LACEY

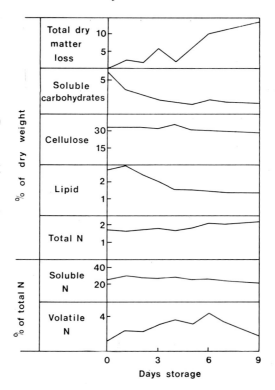

Fig. 7. Biochemical changes in spontaneously heating hay of 40% water content stored in 4·5 l Dewar flasks (plotted from data of Festenstein 1966).

moulding. Conversely, at least at water contents of about 40% when spontaneous heating is greatest, the proportion of nitrogen that is volatile increases as a result of proteolysis and deamination of amino acids. The proportion of nitrogen that is volatile has been correlated with the total number of fungus spores in the hay, with numbers of thermophilic bacteria growing at 60°C in bales, with numbers of fungi growing at 40°C and of actinomycetes and bacteria growing at 60°C in stacks of hay. The lipid content also decreases as hay moulds (Fig. 7; Gregory *et al.* 1963; Festenstein 1966).

pH Value

At baling, the pH of hay is about 6·5. This may fall initially to about 5·5 and then remains within these limits in good hay or in hay that has moulded lightly with fungi of the *Aspergillus glaucus* group, with little spontaneous heating. At higher water contents, with much spontaneous heating, pH may either decrease further to less than 5, perhaps as a result of bacterial action producing acid from carbohydrate, or increase to about 7·5, associated with increased volatile nitrogen, increased fungal colonization and growth of thermophilic actinomycetes. The factors determining the development of acid brown hays or alkaline mouldy hays are imperfectly understood.

Fungi are tolerant of a wide range of pH, most species growing between pH 2 and 8. However, the optimum pH for some thermophilic and thermotolerant species correlates well with their positions in the succession on hay (Table 3). *Mucor miehei* and *M. pusillus* both have pH optima close to the pH of hay at cutting, and both occur early in colonization

TABLE 3

Optimum pH for growth of some thermophilic and thermotolerant fungi (Rosenberg 1975)

Species	Optimum pH
Allescheria terrestris	4·0–6·1
Aspergillus fumigatus	5·9–6·8
Chrysosporum pruinosum	4·1–4·9
Humicola lanuginosa	6·8–7·3
H. stellata	5·4–5·9
Malbranchea pulchella var. *sulfurea*	7·2
Mucor miehei	5·5
M. pusillus	6·1
Talaromyces emersonii	3·4–5·4
T. thermophilus	7·2–8·1

while *Humicola lanuginosa*, *Malbranchea pulchella* var. *sulfurea* and *Talaromyces thermophilus* have optima close to neutrality and occur late in the succession on heated hays. *Aspergillus fumigatus*, a thermotolerant species, is intermediate in its pH requirements, with an optimum pH of 5·9 to 6·8 but growing over the range 2·5 to 8·5. Some fungi characteristic of sugar cane bagasse, where acid conditions are more frequent, e.g. *Allescheria terrestris*, *Chrysosporium pruinosum* and *Talaromyces emersonii*, have pH optima around 5 (Rosenberg 1975).

Actinomycetes and bacteria are mostly less tolerant of acid conditions than fungi and most require a pH between 5·5 and 8·5. The optimum for most actinomycetes is near pH 7 and colonization of hay requires either prior fungal growth or treatment with ammonia.

Other factors

The speed, degree and persistence of heating may be influenced by the size and density of the mass of material. Usually, heating is faster, lasts longer and is more likely to lead to ignition, the larger the stack. Large stacks have a smaller surface area to volume ratio than small. Consequently a smaller proportion forms an inactive mantle of dried material. When the central parts become too hot for further microbial growth, their temperature may be maintained by heating occurring in the outer parts (Miehe 1930; Gregory et al. 1963).

Density is closely connected with oxygen economy. Very dense packing leaving little pore space would prevent spontaneous heating (Miehe 1930) but increasing the density of hay bales from the usual 180 kg m^{-3} to 280 kg m^{-3} results in higher temperatures, perhaps because heat and water losses are decreased (Holden & Sneath 1979).

The Colonization of Hay and Spontaneous Heating

Hay drying in the field becomes colonized by saprophytic 'field' fungi (Christensen & Kaufmann 1974), such as *Alternaria* and *Cladosporium*, which may have been growing as epiphytes in the phylloplane before cutting. Their development is dependent on the prevalent weather conditions, which determine the rate and period of drying necessary before safe storage is possible. They do not usually develop further in storage themselves but it is not known whether they may affect subsequent colonization by typical 'storage' fungi.

An inoculum of storage fungi is also present at cutting, composed of species with differing temperature and water activity requirements. Their numbers may be too small to be detected reliably by usual isolation techniques, but even thermophiles are sufficiently widely distributed to enable uniform moulding of 500 g samples under favourable conditions. Different types of microflora can develop depending on conditions in the bale or stack, of which water content and consequent amount of spontaneous heating are probably most important.

Figure 8 shows the range of water contents and maximum temperatures at which different species of fungi and actinomycetes predominate in hay. The *Aspergillus* spp. are good indicators of the water content at which hay is baled, with *A. glaucus* group, *A. versicolor*, *A. nidulans* and *A. fumigatus* giving maximum numbers with water contents respectively of 25, 29, 31 and 40%. The thermophilic actinomycetes implicated in farmer's lung, *Micropolyspora faeni* and *Thermoactinomyces vulgaris*, only commence growth with 35% water content and heating to about 50°C,

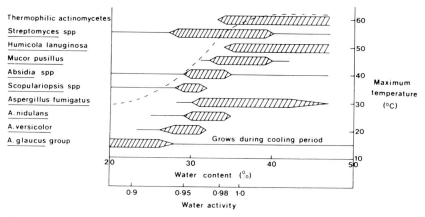

Fig. 8. Water contents and maximum temperatures favouring colonization of hay by different species and groups of fungi and actinomycetes.

reaching maximum numbers in hay with more than 40% water and heating to 60–65°C (Gregory *et al.* 1963; Festenstein *et al.* 1965; Lacey 1978).

Hay baled at 46% water content (Gregory *et al.* 1963) heated to 48°C within 24 h, cooled during the next 24 h to 40°C before reaching 57°C after 3 days and 60°C after 5 days. It then cooled slowly, to 50°C on the 13th day and to 40°C on the 18th. The succession of micro-organisms on the hay can only be deduced by the order in which they sporulate. This could be misleading if species require very different periods of vegetative growth before they produce spores. However, typical 'sugar' fungi were first to appear, *Mucor pusillus* on the second day and *Absidia* spp. on the fourth. These had reached their maxima by the 12th day. Both species have the ability to increase the pH of hay to a level favourable to growth of thermophilic actinomycetes (Pepys *et al.* 1963). These were first isolated from the hay on the 6th day, increasing to a maximum on the 11th. They were accompanied by *Humicola lanuginosa*, considered a secondary sugar fungus (Hedger & Hudson 1974), and *Aspergillus fumigatus*. Both continued to increase until the 16th day. *A. glaucus* group, *A. nidulans* and *Penicillium* spp. only appeared as the hay cooled below 50°C and *Wallemia sebi* only developed during the autumn. Bales with moisture contents less than 46% could mould with *A. glaucus* group from the first day.

Under certain conditions, particularly in stacks of loose hay, an acid bacterial fermentation may occur leading to the formation of brown hay with a characteristic tobacco-like odour and pH less than 5 (Watson & Nash 1960). The conditions leading to this type of fermentation are not fully understood, but may relate to the degree of compression or aeration, the heat and water balance, or the presence of rainwater or dew on the surface of the hay at stacking. However, in an experimental stack, brown hay was interspersed with strata rich in actinomycetes and thermophilic fungi resembling baled hay where there is good aeration (Gregory *et al.* 1963). The bacteria involved in this fermentation have been variously identified as *Bacillus calfactor* (Miehe 1930), *B. licheniformis* (Gregory *et al.* 1963) and *B. stearothermophilus* (Hussain 1973). Acid fermentation and bacterial colonization also appear to be characteristic of hays that heat to ignition temperatures (Festenstein 1971; Hussain 1973).

Colonization of Other Substrates

Colonization of other organic substrates associated with spontaneous heating appears to follow patterns similar to that of hay, although the species involved may be different. For instance, *M. faeni* and *T. vulgaris* are replaced in sugar-cane bagasse by *Thermoactinomyces sacchari*, *Saccharopolyspora hirsuta* and *Pseudonocardia* spp.; in mushroom compost *Thermomonospora* spp. and *Scytalidium thermophilum* have predominated; in grain, *Actinomadura*-like isolates are often common, and *Penicillium* spp. are more numerous than in hay (Lacey 1973, 1978). For most substrates, knowledge of the microflora is incomplete because workers have identified only the fungi, actinomycetes or bacteria.

Applied Aspects of Spontaneous Heating

Processes involving spontaneous heating are important in the flavouring of some foods, the preparation of composts and disposal of waste. Thermophilic fungi have also been implicated in the retting process to improve latex extraction from guayule.

Food fermentations, usually based on soybeans or rice, are a feature of oriental cookery and also occur in South America (Tansey & Brock 1978). Pure cultures of fungi are often used and the process may involve spontaneous heating. For example, in the preparation of miso by fermenting rice with *Aspergillus oryzae*, the temperature must be controlled below 40°C (Hesseltine 1965). Fermentations are also essential to the flavouring of cocoa, coffee and black teas. However, *M. pusillus* and *A. fumigatus* may develop abundantly on poor cocoa fermentations (Broadbent & Oyeniran 1968).

Microbial colonization is essential for the disposal of organic waste in nature. In agriculture and horticulture, it is utilized in composting where ideally heating should be sufficient to kill weed seeds; in the preparation of mushroom composts from horse manure and straw; and the preparation of hot-beds for the cultivation of some crops. A similar process is used by the Mallee fowl to incubate its eggs. The temperature of the composting material around the eggs is controlled within narrow limits by the bird using its beak as a temperature probe (Frith 1962).

Recently, the use of thermophilic actinomycetes in the composting of municipal waste has provoked much interest (Stutzenberger et al. 1970; Kane & Mullins 1973).

Consequences of Microbial Colonization and Spontaneous Heating

The microbial colonization and heating of stored materials can be lead to losses of quality and nutritional value and also to potential health hazards to workers handling the materials and to people and animals consuming them.

Quality of cereal grain for seed or malting may be decreased through loss of viability while baking qualities are adversely affected by heating above 40°C (Westermarck-Rosendahl 1978). Nutritional value is lost through loss of dry matter, destruction of major, minor and trace nutrient, and loss of digestibility of proteins (Watson & Nash 1960; Lacey 1975a). Total loss may occur if heating leads to spontaneous ignition.

Possible health hazards include infection, allergy and mycotoxicosis. *Aspergillus fumigatus* is well known as a pathogen causing mycotic abortion of cattle and respiratory infections of man and animals. *Absidia* spp. may also be implicated in mycotic abortion and like *Mucor pusillus* may cause phycomycosis. Allergies of two types may occur to fungus and actinomycete spores. Type 1 allergy occurs immediately on exposure to a sensitized person to the allergen, giving hay fever-like symptoms and asthma, while a Type 3 allergic reaction occurs several hours after exposure, affects the gas-exchange tissue of the lung and is characteristically occupational. *Aspergillus fumigatus* is a common cause of asthma while several thermophilic actinomycetes have been implicated in allergic alveolitis. Toxic metabolites are produced by many fungi but those of *Aspergillus* and *Penicillium* spp. that are often common in food and feedstuffs are best known (Lacey et al. 1972; Lacey 1975a,b).

References

AYERST, G. 1965 Water activity—its measurement and significance in biology. *International Biodeterioration Bulletin* **1**, (2), 13–26.

AYERST, G. 1969 The effects of moisture and temperature on growth and spore germination in some fungi. *Journal of Stored Products Research* **5**, 127–141.

BROADBENT, J. A. & OYENIRAN, J. O. 1968 A new look at mouldy cocoa. In *Biodeterioration of Materials* eds Walter, A. H. & Elphick, J. J. London: Elsevier Publishing Co. Ltd.

CARTER, E. P. & YOUNG, G. Y. 1950 Role of fungi in the heating of moist wheat. *United States Department of Agriculture Circular No. 838.*

CHRISTENSEN, C. M. & KAUFMANN, H. H. 1974 Microflora. In *Storage of Cereal Grains and their Products* ed. Christensen C. M. St. Paul: American Association of Cereal Chemists.

COONEY, D. G. & EMERSON, R. 1964 *Thermophilic Fungi.* San Francisco: W. H. Freeman & Co.

CROSS, T. 1968 Thermophilic actinomycetes. *Journal of Applied Bacteriology* **31**, 36–53.

CURRIE, J. A. & FESTENSTEIN, G. N. 1971 Factors defining spontaneous heating and ignition of hay. *Journal of the Science of Food and Agriculture* **22**, 223–320.

DEPLOEY, J. J. & FERGUS, C. L. 1975 Growth and sporulation of thermophilic fungi and actinomycetes in O_2-N_2 atmospheres. *Mycologia* **67**, 780–797.

DYE, M. H. 1964 Self-heating of damp wool. Part 1. The estimation of microbial populations in wool. *New Zealand Journal of Science* **7**, 87–96.

DYE, M. H. & ROTHBAUM, H. P. 1964 Self-heating of damp wool. Part 2. Self-heating of damp wool under adiabatic conditions. *New Zealand Journal of Science* **7**, 97–118.

EASSON, D. L. & NASH, M. J. 1978 Preservation of moist hay in miniature bales treated with propionic acid. *Journal of Stored Products Research* **14**, 25–33.

FERGUS, C. L. 1964 Thermophilic and thermotolerant molds and actinomycetes of mushroom compost during peak-heating. *Mycologia* **56**, 267–284.

FERGUS, C. L. 1969*a* The cellulolytic activity of thermophilic fungi and actino-mycetes. *Mycologia* **61**, 120–129.

FERGUS, C. L. 1969*b* The production of amylase by some thermophilic fungi. *Mycologia* **61**, 1171–1175.

FESTENSTEIN, G. N. 1966 Biochemical changes during moulding of self-heated hay in Dewar flasks. *Journal of the Science of Food and Agriculture* **17**, 130–133.

FESTENSTEIN, G. N. 1971 Carbohydrates in hay on self-heating to ignition. *Journal of the Science of Food and Agriculture* **22**, 231–234.

FESTENSTEIN, G. N., LACEY, J., SKINNER, F. A., JENKINS, P. A. & PEPYS, J. 1965 Self-heating of hay and grain in Dewar flasks and the development of farmer's lung antigens. *Journal of General Microbiology* **41**, 389–407.

FRITH, H. L. 1962 *The Mallee-fowl. The Bird that builds an Incubator.* Sydney: Angus Robertson Ltd.

GREGORY, P. H. & LACEY, M. E. 1963 Mycological examination of dust from mouldy hay associated with farmer's lung disease. *Journal of General Microbiology* **30**, 75–88.

GREGORY, P. H., LACEY, M. E., FESTENSTEIN, G. N. & SKINNER, F. A. 1963 Microbial and biochemical changes during the moulding of hay. *Journal of General Microbiology* **33**, 147–174.

HEDGER, J. N. & HUDSON, H. J. 1974 Nutritional studies of *Thermomyces lanuginosus* from wheat straw compost. *Transactions of the British Mycological Society* **62**, 129–143.

HENSSEN, A. 1957a Beiträge zur Morphologie und Systematik der thermophilen Actinomyceten. *Archiv für Mikrobiologie* **26**, 373–414.

HENSSEN, A. 1957b Über die Bedeutung der thermophilen Mikroorganismen für die Zersetzung der Stallmistes. *Archiv für Mikrobiologie* **27**, 63–81.

HESSELTINE, C. W. 1965 A millennium of fungi, food and fermentation. *Mycologia* **57**, 149–197.

HOLDEN, M. R. & SNEATH, R. W. 1980 Performance of preservatives in conventional-sized bales. *NIAE Departmental Note* DN/M /1365 (in preparation).

HUSSAIN, H. M. 1973 Ökologische Untersuchungen über die Bedeutung thermophiler Mikroorganismen für die Selbsterhitzung von Heu. *Zeitschrift für Allgemeine Mikrobiologie* **13**, 323–334.

JAMES, L. H. 1927 Studies in microbial thermogenesis. I. Apparatus. *Science* **65**, 504–506.

JAMES, L. H., RETTGER, L. F. & THORN, C. 1928 Microbial thermogenesis II. Heat production in moist organic materials with special reference to the part played by micro-organisms. *Journal of Bacteriology* **15**, 117–141.

JAMES, N. & LEJEUNE, A. R. 1952 Microflora and the heating of damp stored wheat. *Canadian Journal of Botany* **30**, 1–8.

KANE, B. E. & MULLINS, J. T. 1973 Thermophilic fungi in a municipal waste compost system. *Mycologia* **65**, 1087–1100.

KUO, M. J. & HARTMAN, P. A. 1966 Isolation of amylolytic strains of *Thermoactinomyces vulgaris* and production of thermophilic actinomycete amylases. *Journal of Bacteriology* **92**, 723–726.

LACEY, J. 1971 The microbiology of moist barley storage in unsealed silos. *Annals of Applied Biology* **69**, 187–212.

LACEY, J. 1973 Actinomycetes in soils, composts and fodders. In *Actinomycetales: Characteristics and Practical Importance* eds Skinner, F. A. & Sykes, G. *Society for Applied Bacteriology Symposium Series* No. 2, pp. 231–251, London: Academic Press.

LACEY, J. 1974 Moulding of sugar-cane bagasse and its prevention. *Annals of Applied Biology* **76**, 63–76.

LACEY, J. 1975a Potential hazards to animals and man from micro-organisms in fodders and grain. *Transactions of the British Mycological Society* **65**, 171–184.

LACEY, J. 1975b Occupational and environmental factors in allergy. In *Allergy '74* eds Ganderton, M. A. & Frankland, A. W. pp. 303–319, London: Pitman Medical.

LACEY, J. 1978 The ecology of actinomycetes in fodders and related substrates. In *Nocardia and Streptomyces* eds Mordarski, M., Kurylowicz, W. & Jeljaszewicz, J. pp. 161–170, Stuttgart: Gustav Fischer Verlag.

LACEY, J. & DUTKIEWICZ, J. 1976a Methods for examining the microflora of mouldy hay. *Journal of Applied Bacteriology* **41**, 13–27.

LACEY, J. & DUTKIEWICZ, J. 1976b Isolation of actinomycetes and fungi using a sedimentation chamber. *Journal of Applied Bacteriology* **41**, 315–319.

LACEY, J., PEPYS, J. & CROSS, T. 1972 Actinomycete and fungus spores in air as respiratory allergens. In *Safety in Microbiology* eds Shapton, D. A. & Board, R. G. *Society for Applied Bacteriology Technical Series* No. 6, pp. 151–184, London: Academic Press.

MIEHE, H. 1930 Über die Selbsterhitzung des Heues. 2 Auflage. *Arbeiten der Deutschen Landwirtschafts—Gesellschaft* **196**, 1–47.

OSO, B. A. 1978 The production of cellulase by *Talaromyces emersonii*. *Mycologia* **70**, 577–585.

PEPYS, J., JENKINS, P. A., FESTENSTEIN, G. N., GREGORY, P. H., LACEY, M. E. & SKINNER, F. A. 1963 Farmer's lung: thermophilic actinomycetes as a source of "farmer's lung hay" antigen. *Lancet* **2**, 607–611.

PITT, J. I. & CHRISTIAN, J. H. B. 1968 Water relations of xerophilic fungi isolated from prunes. *Applied Microbiology* **16**, 1853–1858.

RAMSTAD, P. E. & GEDDES, W. F. 1942 The respiration and storage behaviour of soybeans. *Minnesota Agricultural Experimental Station Technical Bulletin* No. 156.

ROSENBERG, S. L. 1975 Temperature and pH optima for 21 species of thermophilic and thermotolerant fungi. *Canadian Journal of Microbiology* **21**, 1535–1540.

ROSENBERG, S. L. 1978 Cellulose and lignocellulose degradation by thermophilic and thermotolerant fungi. *Mycologia* **70**, 1–13.

ROTHBAUM, H. P. & DYE, M. H. 1964 Self-heating of damp wool Part 3. Self-heating of damp wool under isothermal conditions. *New Zealand Journal of Science* **7**, 119–146.

SINHA, R. N. & WALLACE, H. A. H. 1965 Ecology of a fungus-induced hot spot in stored grain. *Canadian Journal of Plant Science* **45**, 48–59.

SMITH, R. S. & OFOSU-ASIEDU, A. 1972 Distribution of thermophilic and thermotolerant fungi in a spruce-pine chip pile. *Canadian Journal of Forest Research* **22**, 16–26.

SNOW, D., CRICHTON, M. H. G. & WRIGHT, N. C. 1944 Mould deterioration of feeding-stuffs in relation to humidity of storage. Part II. The water uptake of feeding-stuffs at different humidities. *Annals of Applied Biology* **31**, 111–116.

STUTZENBERGER, F. J., KAUFMAN, A. J. & LOSSIN, R. D. 1970 Cellulolytic activity in municipal solid waste composting. *Canadian Journal of Microbiology* **16**, 553–560.

TANSEY, M. R. 1971 Isolation of thermophilic fungi from self-heated, industrial wood-chip piles. *Mycologia* **63**, 537–547.

TANSEY, M. R. & BROCK, T. D. 1978 Microbial life at high temperatures: ecological aspects. In *Microbial Life in Extreme Environments*, ed Kushner, D. J. London: Academic Press.

WAITE, R. 1949 The relationship between moisture content and moulding in cured hay. *Annals of Applied Biology* **36**, 496–503.

WALKER, I. K. & HARRISON, W. J. 1960 The self-heating of wet wool. *New Zealand Journal of Agricultural Research* **3**, 861–895.

WALKER, I. K. & WILLIAMSON, H. M. 1957 The spontaneous ignition of wool. I. The causes of spontaneous fires in New Zealand Wool. *Journal of Applied Chemistry* **7**, 468–480.

WATSON, S. J. & NASH, M. J. 1960 *The Conservation of Grass and Forage Crops.* 2nd edn, Edinburgh: Oliver & Boyd.

WESTERMARCK-ROSENDAHL, C. 1978 Spontaneous heating in newly harvested wheat and rye. II. Effects on the technological quality of flour. *Acta Agriculturae Scandinavica* **28**, 159–168.

WESTERMARCK-ROSENDAHL, C. & YLIMÄKI, A. 1978 Spontaneous heating in newly harvested wheat and rye. I. Thermogenesis and its effect on grain quality. *Acta Agriculturae Scandinavica* **28**, 151–158.

The Growth of Microbes at Low pH Values

M. H. Brown and T. Mayes

Unilever Research Laboratory, Colworth House, Sharnbrook, Bedford, UK

H. L. M. Lelieveld

Unilever Research Laboratory, Vlaardingen, The Netherlands

Acidity is widely used in food preservation, often synergistically with a wide range of other factors including reduced a_w, *Eh*, low temperature or lowered oxygen tension; it is also used as an aid in canning processes (Mossel 1971). However, many microbes are able to survive and grow in acid environments, although in some cases the acidity of the environments (pH less than 3) is sufficient to denature or at least inhibit the activity of enzymes and other acid-labile molecules such as ATP and DNA which are essential to the organism (Brock 1969).

The hydrogen ion concentration affects the ionic state and therefore the availability of many metabolites and inorganic ions to an organism, hence its influence on the state and activity of many cellular components cannot be over emphasized. The fact that many organisms are able to grow at extreme pH values suggests that they have some mechanism for maintaining the internal pH near to normal physiological values. Although pH is used commonly as an index of acidity, when applied to a single microbial cell, having a volume of about 5×10^{-14} ml and containing (on a statistical basis) 3·6 free protons at pH 7, its usefulness is uncertain as it does not take account of any undissociated protons bound to donor molecules or the average degree of ionization of charged groups within the cell (Langworthy 1978).

Although microbial cells need to maintain a tolerable internal pH value, other factors determine the range of pH values which will permit growth. The work of Mitchell and others (see Hamilton 1975) has suggested that, in order to grow, microbes must maintain a regulated flow of protons (H^+) across their membranes. Membrane-bound enzymes mediate or regulate the size and variation of this flow, which is used for nutrient transport and may under certain circumstances provide energy (Garland 1977). Under adverse conditions, at acid external pH values, the energy yielding processes situated in the cell membrane may be reversed and used to pump protons from the cell's internal space, so that enzyme reactions can proceed under favourable (neutral pH) conditions. We intend to set out simple methods which may be used to study the influence of pH (or for that matter any other stress factor) on the way in which microbial cells use energy to counter the effects of hostile environments. The effects of environmental stress on cellular energetics have not been studied extensively, although the regulation of two factors: (1) the proton flow between the cell interior and the environment and (2) the internal H^+ concentration are both essential to growth. The balancing of these two factors must play a major role in determining the economy of the cell and hence the influence of pH is an ideal subject for study using quantitative techniques to study cellular economy. A simplified view of the cellular functions associated with the membrane of *Streptococcus faecalis* is shown in Fig. 1.

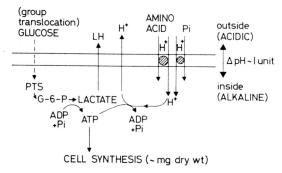

Fig. 1. A simplified view of the cellular functions associated with the cell membrane of *S. faecalis* that are relevant to growth at low pH. PTS, phosphotransferase transport system; G-6-P, glucose-6-phosphate; LH, undissociated lactic acid; Pi, inorganic phosphate; ΔpH, transmembrane pH gradient; ADP, adenosine diphosphate; ATP, adenosine triphosphate.

Batch Culture

pH tolerance, growth rate and yield

Measurement

Tolerance to pH is easily measured by growth or no growth in a particular medium; however it is a matter of common experience that when organisms grow under stress their growth rate and final concentration may vary according to the conditions around them (Fig. 2). Specific growth rate (μ, increase in the number of cells per hour) and molar growth yield (Y_g, mg dry weight of cells produced per mole of substrate utilized) are two expressions which may be conveniently used to qualify the growth of microbes (Stouthamer & Bettenhaussen 1973). We have used optical density (D) measurements of small scale (3·5–20 ml) batch cultures as a means of following the growth of *S. faecalis* cultures and estimating μ and Y_g (Fig. 2). To work over a range of pH values requiring many samples it is impractical to measure growth in any other way.

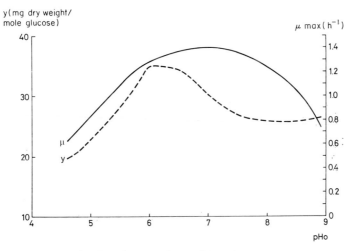

Fig. 2. The influence of pH on the growth rate (μ) and yield (y) of *S. faecalis* grown in batch culture at 37°C. pHo is initial pH.

However it is not always possible to use optical density to obtain a true estimate of growth as the morphology of certain organisms destroys the linear relationship often assumed between D and cells ml^{-1}. Sometimes, the weight of organisms grown on one substrate may be accurately quantified using D, whereas when they are grown on another substrate no relationship can be found. When *S. faecalis* is grown on a rich medium containing glucose as the energy source, turbidity is related to dry weight of cell material (established by drying to constant weight), $D = f$ (dry weight; Pirt 1975); but if the same organism is grown with arginine + galactose as the energy source, then such a relationship is not found. In a rich medium *S. faecalis* used over 90% of the added glucose as an energy source (Bauchop & Elsden 1960).

As the geometry of sample tubes and spectrophotometers varies there is no universal relationship between D and mg dry wt of cells ml^{-1}, even when sample containers with the same optical path length are used (see Mallette 1970). We have used plastic disposable cuvettes ($10 \times 10 \times 40$ mm; Sarstedt Cat. No. 740) containing 3·5 ml of culture medium for some work, but more often we have used colorimeter tubes (15 mm ϕ, Corning) containing 10 ml, as they are optically matched and, being cylindrical, can be easily incubated in a dry heating block with drilled holes of similar diameter (Scientia & Cook Electronics Ltd., London, England). Incubation of the tubes in a water bath introduced unnecessary errors as repeated wetting and drying led to a build-up of opaque substances on the outside of the tube.

It is important that the cultures can be accurately maintained at a known temperature over the period of the experiment, fluctuations greater than $\pm 0\cdot1°C$ can affect growth rate significantly and may lead to erroneous interpretation of results.

Growth should be related to a blank, preferably of culture medium, but this poses problems for experiments extending over long periods as the blank may become contaminated and growth may occur if it is kept at room temperature. The blank may be kept refrigerated, but care should be taken to ensure that the walls of the tube are free of condensation during the period it is used for calibration; alternatively an adequate supply of the same medium may be kept chilled and dispensed as needed. We used either a Hilger H810-1 or a Vitatron MPS spectrophotometer to measure the absorbance of the broths. As the culture medium used was brown, the light source used was 520–580 nm, as this range of wavelengths is minimally absorbed by the uninoculated medium.

Light scattering (nephelometry) may be used for the determination of low turbidity levels (equivalent to about 10^5 organisms ml^{-1} in cuvettes with a 1 cm light path), but in culture medium prepared without special attention being paid to suspended solids, the true signal (microbial growth) to noise ratio is often unacceptably high and variable.

Growth rate

We measured growth rate (μ) by following the increase in optical density over the whole period of growth from lag phase to stationary phase; maximum growth rate (μ_{max}) was only found to span a small portion of the whole exponential phase of growth. Hence, during some of the exponential phase of growth the organisms divided at a constant rate and, if x is the concentration of organisms (mg dry weight ml^{-1}), then μ_{max} is given by the equation:

$$\mu_{max} = \frac{1}{x} = \frac{dx}{dt}$$

The D of cultures growing at various pH values was measured at short intervals (10–30 min) and the data acquired were processed using a computer program.

The program used calculated growth rates (μ) by fitting cubic splines through the natural logarithm of the measured D value (ln D) in such a way that the curve endings of the calculated splines joined smoothly at their junctions (see Fig. 3). Such treatment smoothed statistically the lines derived from the original data (Fig. 4). This smoothing is necessary because of the small sample size (3–20 ml) used in this work. The computer program produced the first derivative of the curves d ln D/dt to

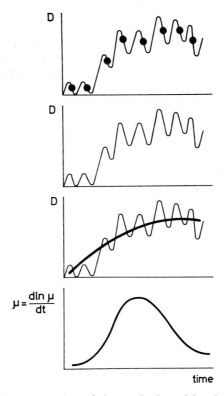

Fig. 3. A graphical representation of the method used by the cubic spline program to generate smooth S-shaped curves from data collected during experimental work. This method of data smoothing is useful when the original signal to noise ratio is low. The figures show the experimental data points (●) which the computer links with S-shaped curves (cubic splines), the data point lying at the centre of the spline. The splines derived from adjacent points are joined so that the arms of the splines link with no difference in gradient. This junction is used to provide an interpolated point which is in turn used to generate another spline. Up to 50 interpolated points may be generated from each data point and these are used by the computer to calculate the line of best fit. (D = optical density.)

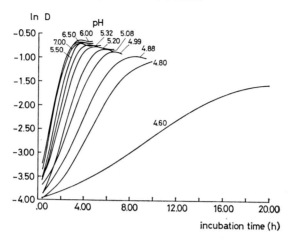

Fig. 4. The lines derived from the data points recorded at intervals (30 min) during the growth of *S. faecalis* in broth buffered to the pH values indicated on the graph.

calculate the specific growth rates at any time (Fig. 5) and also the time taken to reach μ_{max} (Fig. 6). Because $x \alpha D$ the equation becomes:

$$\mu - \frac{d \ln k \times D}{dt}$$

(when k is a constant) therefore:

$$\mu = \frac{d \ln D}{dt} \times \frac{dk}{dt}$$

and thus:

$$\mu = \frac{d \ln D.}{dt}$$

Yield

Growth yields (Y) were found by observing the final D of cultures. For meaningful results the energy source must be the limiting nutrient and it must be known whether the culture is aerobic or anaerobic. It is difficult to ensure satisfactory aeration of cultures to determine Y_{O_2}, as gaseous exchange and oxygen demand values will be dependent on growth rate and culture density. Although *S. faecalis* is a fermentative organism, more reproducible results were obtained in our experiments when cultures were flushed with nitrogen and closed at the start of the experiment.

Molar growth yields can only be calculated if the consumption of glucose or other energy source during growth is known. We calculated molar growth yield values for glucose (Y_g mole glucose $^{-1}$) by using a

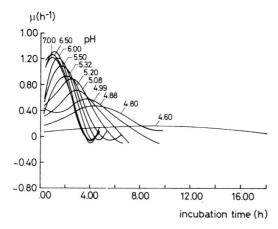

Fig. 5. The first derivative d ln D/dt of the curves shown in Fig. 4; these curves indicate the maximum growth rate, the time for the culture to reach maximum growth rate (μ_{max}) and the period of active growth.

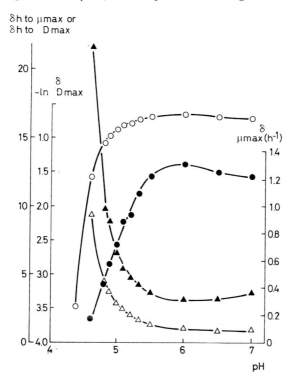

Fig. 6. A summary of the information derived from one set of data and used to define microbial growth in response to different pH values. δ to μ_{max} ($\triangle$); δ to D_{max} ($\blacktriangle$); —ln D_{max} ($\circ$); μ_{max} h^{-1} ($\bullet$) D = optical density.)

measured conversion value to relate the D to dry weight values: the value of this factor was 1 D unit $= 300$ mg dry wt cells 1^{-1} (see Fig. 2). The weights obtained in this way were related to glucose consumption which was measured using the GOD-Perid method (Boehringer, Mannheim GmbH).

Samples of the culture liquid (0·2 ml) were mixed with uranyl acetate (2 ml; 0·16 % w/v in saline, 0·9 % w/v) to precipitate the cells. This supernatant was freed of cells by centrifuging ($>1200 \, g$) and it was incubated with the enzyme/indicator solution at 25°C for 25 min. The colour development was measured in a spectrophotometer at 610 nm using a blank and control for calibration.

It is only possible to estimate the influence of stress on yield if the energy source is growth-limiting. For this reason it is desirable to use low concentrations of energy source (e.g. about 5 mM).

We have only estimated $Y_{glucose}$, although Y_{ATP} is known to provide a more accurate picture of cellular economy (see Stouthamer 1970) because in order to convert $Y_{glucose}$ to Y_{ATP} a complete analysis of fermentation end-products is needed (so that the potential yield of ATP is accurately known). When working with *S. faecalis* it is important to take the highest observed D value as the yield, because soon after exhaustion of the energy source the D falls and readings taken even 30 min after this point will suggest a low yield (Fig. 4). Moustafa & Collins (1968) found a similar result with other lactic acid bacteria and presumably it is true of other organisms.

Problems of batch cultures

As conditions within a batch culture change rapidly during growth, it is not possible to estimate the influence of particular environmental pH values on the maximum growth rate (μ_{max}). It is particularly difficult to estimate the influence of pH on the growth of an acid producing microbe such as *S. faecalis*. With acid producing microbes, the maintenance of a constant pH in the medium is a major difficulty. It cannot be achieved without altering the ionic composition of the medium by the addition of alkali. Hence the changing environment in a batch culture limits the extent to which information derived in this way can be related to the metabolic activities of microbes.

Medium and buffers

In this work the complex culture medium: tryptone (Difco), 2 g 1^{-1}; yeast extract (Difco), 1 g 1^{-1}; glucose, 1 g 1^{-1} was buffered to limit the pH shift during growth. The pH was adjusted initially with HCl (1 M) or NaOH (2 M) to an accuracy of ±0.01 pH units. Either histidine (0·1 M; BDH Biochemicals 37221; pH range 4–7) or HEPES (Good *et al.* 1966; *N*-2-hydroxyethylpiperazine-*N*-2-ethane sulfonic acid, 0·1 M; Sigma Chemicals, Poole, Dorset No. H3375; pH 7–9) was included in the medium as a buffer. These two buffers were chosen as, at the concentration used (0·1 M) neither served as an energy source nor inhibited growth. At this concentration in a medium containing glucose (1 mg ml^{-1}) the pH had decreased by less than 0·4 units in the histidine-buffered range and less than 0·6 units in the HEPES-buffered range when the glucose was completely consumed.

This glucose concentration (5·5 mM) gave sufficient growth for us to follow accurately by measurement of optical density, whilst the amount of acid produced from it was not too great to be accommodated by the buffer at a concentration which did not interfere unacceptably with growth. Higher buffer concentrations could reduce pH variation, but significantly influenced growth. It is possible that, in batch culture, the accumulation of inhibitory metabolites also influenced growth; their nature and quantity will be influenced by environmental pH, growth rate and nutrient availability.

Analysis of results

The results were analysed so that μ_{max} and $Y_{glucose}$ could be estimated. We processed batch growth data using Harwell subroutine library methods VC 03A/AD (Sept, 1972) and VB 06A/AD (Dec, 1973) (Theoretical Physics Division, UKAEA Research Group, Atomic Energy Research Establishment, Harwell) to smooth the growth curves (Figs 4, 5 & 6) and provide data on the maximum growth rate by calculating the first derivative. The results were displayed graphically using a Harris graph plotter.

The purpose of VB 06A/AD is to calculate a weighted least squares fit to given data values by a cubic spline which has knots and smoothing factors assigned by the user; VC 03A/AD calculates a smooth curve fit to given weighted data. The output is in a form which allows easy evaluation of the fitted junction at any point in the range of the data, so that the first derivative (μ_{max}) can be found (see Figs 4, 5 & 6).

Use of inhibitors to investigate mechanisms of resistance

Metabolic inhibitors which have a known mode of action may be used to identify particular mechanisms important in microbial resistance to stress. It is attractive to speculate that an assortment of chemicals which

are broadly categorized in the same way, for example as proton ionophores or ATPase inhibitors, will interact according to their gross effects.
However, it is unlikely that this is so as these chemicals have specific
sites of action, presumably interfering with different macromolecular
mechanisms. Several functional classes of compound are likely to be of
special importance in determining the mechanisms used by microbes to
resist the effects of low pH.

Proton ionophores

Proton ionophores make the normally impermeable microbial membrane freely permeable to protons. In sufficient quantity they should
allow the pH within the cell to equilibrate with that of the environment.
At pH values permitting growth, the presence of ionophores could increase the energy demands of the organism (reducing yield) as the action
of membrane-bound ATPase would be reversed (consuming ATP) and
used for pumping protons out of the cell. Removal or reduction of the
proton gradient could also limit nutrient transport and would hence
reduce growth rate; under such circumstances it is important to ensure
that glucose remains the limiting nutrient. At low pH values essential
enzymes may be inactivated if the pH gradient (inside of the cell usually
more alkaline than the outside environment) is removed; proton ionophores may reduce the pH tolerance of microbes. Under certain circumstances the addition of proton ionophores, such as tetrachlorsalicylanilide
(TCS, Kodak Limited, Kirkby, Liverpool) may cause the uncoupling of
glucose consumption from growth (see Fig. 7; Harold & Baarda 1968). It
has been suggested that energy is diverted from biosynthesis into either
'slip reactions' or 'futile cycles' (Tempest 1978) or to satisfy an increased
demand for maintenance energy which is used to maintain concentration
gradients between the cell and its exterior (Pirt 1975). The balance for
energy source utilization is given by:

$$\text{total rate of consumption} = \text{rate of consumption for growth} + \text{rate of consumption for maintenance and by slip reactions}$$

ATP may be generated, or lost, through its hydrolysis not being coupled
to growth or maintenance functions; such lost ATP would normally be
indistinguishable from maintenance energy. Gomez-Puyou & Gomez-
Lojero (1977) surveyed the use of ionophores to study the function of
biological membranes.

ATPase inhibitors

If microbes alter their rate of energy transfer, or the balance between
usage for biosynthesis and uses elsewhere, in response to adverse

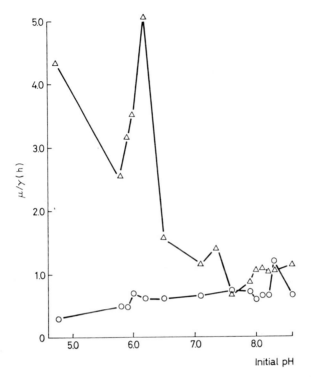

Fig. 7. The specific growth rate (μ) divided by the specific glucose uptake rate (γ) at various pH values before ($\bigcirc$) and after ($\triangle$) the addition of tetrachlorosalycilanilide (TCS, 5×10^{-5} M). Around pH 6, after the addition of TCS, the rate of glucose uptake is much greater than the rate of growth, indicating uncoupling.

environments then inhibition of ATPase activity should reduce pH tolerance and growth rate. Under adverse conditions the inhibition of ATPase activity should lead to a reduction in the rate at which energy can be made available for maintenance of the cell's integrity. For this reason when suitable ATPase inhibitors are present, yield may be increased, as energy cannot be used sufficiently rapidly to counter stress factors, such as low pH, effectively. As a result only growth rate and pH tolerance would be affected because the interior of the cell is acidified more easily than when ATPases are fully functional. Dio-9, an antibotic of unknown structure, and the synthetic bis-guanidine chlorhexidine (1-6-di-4'-chlorophenyldiguanidohexane) are thought to inhibit ATPase activity when added to intact cells of *S. faecalis* (Harold *et al.* 1969). When added to these cells Dio-9 (7 μg ml^{-1}; Gist-Brocades; P.O. Box 1. Wateringseweg 1, Delft, The Netherlands) and chlorhexidine (2 μg ml^{-1};

ICI Ltd., Pharmaceuticals Division, Wilmslow, Cheshire, UK) inhibited the net uptake of K^+ by exchange for Na^+ and H^+.

Effects of inhibitors

Other inhibitors may be useful in outlining the mechanisms of resistance if their action is on one of the primary mechanisms of resistance used by the organisms; their action may also be investigated by measuring stress-tolerance, growth rate and yield. But often, besides their primary effect, inhibitors will exert a range of secondary effects, associated with the specific chemical groups that they modify or bind to, for example, some ionophores attack disulphide bonds. Hence an ionophore which lodges in the membrane, to form an ion channel, may also interfere with other metabolic activities in the membrane besides conducting ions between the cell interior and the environment. The importance of these activities in the microbe's tolerance of stress is unknown, as primary effects and consequences cannot be identified easily.

One way of avoiding the pitfalls of inhibitor studies is to use mutants which have well characterized deficiencies in the biochemical activities being studied (see below).

Many of the inhibitors used in these studies are substantive (i.e. bind to cellular components), and for this reason the inoculum or population level relative to the concentration of inhibitor needs to be known. If a small inoculum is used the cells are exposed to a much higher effective level than if a large one is added. *S. faecalis* was more sensitive to ionophores in continuous than in batch culture. Some of the inhibitors we have used gave a dramatic increase in inhibition for a small increase in level; for example DCCD (N, N'-dicyclohexylcarbodiimide, Sigma Chemicals, London) gave 20% inhibition of growth rate at $1 \mu g$ ml^{-1} (pH 5·6) but 80% inhibition at $2 \mu g$ ml^{-1} (pH 5·6) and only 90% at $10 \mu g$ ml^{-1} (pH 5·6).

Standard growth method

Cultures used to determine the interaction between pH and inhibitors were grown under carefully standardized conditions to ensure that the cells remained as nearly comparable as possible throughout all of the experiments. In matrix experiments done to perm 8 pH values (pH 4·8–8·5) with a number of inhibitors the reliability of the standard growth method (designed to produce comparable numbers of cells in a similar metabolic state) was checked by including some combinations from previous runs, as controls in each experiment.

Streptococcus faecalis was grown in static culture (18 h$\pm$15 min at 37°C); 0·05 ml was used to inoculate fresh medium (250 ml) and this was incubated for $18\,h \pm 15$ min at $37°C$. The cells were harvested by centrifugation and washed twice in chilled ($4°C$) $MgCl_2$ solution (2 mM). The culture medium used was $10 \times$ strength basal medium, it was sterilized by autoclaving at $121°C$ for 15 min. The glucose and histidine buffer ($0·1$ M) were autoclaved separately and added aseptically; the pH was corrected to $6·8$ after autoclaving.

The basal medium (TGY) used in all the experiments contained (per litre): tryptone (Difco), 2 g; yeast extract (Difco), 1 g; glucose, 1 g; histidine or HEPES, $0·1$ M. Low levels of nutrients were used as *S. faecalis* can co-metabolize the arginine, present in tryptone and yeast extract, with glucose to yield energy (Bauchop & Elsden 1960). We calculated that the amount of arginine included in the medium was sufficient to produce $0·05$ mM ATP by co-metabolism, whereas there was enough glucose to produce $1·1$ mM ATP.

Assessing the effects of inhibitors

We found that the most reliable way to assess the influence of inhibitors on growth rate was to grow the culture in the basal medium at a particular pH value until sufficient data points (from D measurements) were available to give a good estimate of μ_{max} (exponential growth phase straight line, correlation coefficient $>0·9$, using 5 data points: the gradient and correlation coefficient were calculated using a Texas programmable calculator Model SR-56). At this point the culture was diluted by half with fresh basal medium of the same pH value, containing twice the required concentration of inhibitor. The growth rate in the modified medium was then found in the same way. We have used this method to demonstrate the uncoupling of glucose uptake from growth in the presence of TCS (Fig. 7). The uncoupling of glucose uptake from growth may be shown (Fig. 7) by using a dimensionless plot; i.e. when the specific glucose uptake rate a d $\ln C/dt$ is divided by the specific growth rate $\mu = $ d $\ln x/dt$ the result should be one, when coupling is complete. When uncoupling occurs the rate of glucose uptake will exceed the growth rate.

pH–shocking experiments

When the environment surrounding a microbe is changed abruptly an adaptive response is often evoked. Gould & Measures (1977) have found that a shift in environmental water activity from $0·995$ to $0·97$ caused an adaptive response in *Bacillus subtilis*. The length of this period of adaption could be found using D (Fig. 8). We have investigated the effect of an abrupt change in environmental pH on the growth of *E. coli* and *S. faecalis* (Fig. 8). Cultures of *S. faecalis* were grown in the basal

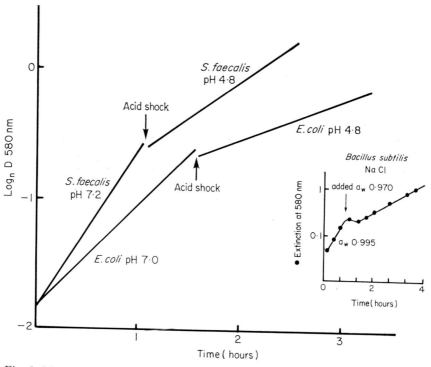

Fig. 8. The effect of acid-shocking *S. faecalis* and *E. coli* from neutral pH to pH 4·8. The inset shows the adaptive response of *B. subtilis*, when culture water activity (a_w) is reduced from 0·995 to 0·970. (Data from Gould & Measures 1977.)

medium (TGY) until their μ_{max} had been established by D measurement and calculation, the pH of this medium was kept constant by the addition of alkali. It could not be buffered as this would make the rapid pH shift difficult to achieve. A sample (10 ml) of the culture (250 ml) was then filtered rapidly through a Millipore filter (MAWP 02500 MA 0·45 μ pore size; Fig. 9). The filter membrane together with the cells adhering to it was then aseptically placed into fresh medium (10 ml) at the lowered pH and the new growth rate determined. Using this method cells could be rapidly shocked without a substantial decrease in optical density, the use of fresh medium minimized the influence of end-product accumulation. For pH-shift experiments on *E. coli* a modified Spizizens broth [g l^{-1}: (NH$_4$)$_2$SO$_4$, 2; K$_2$HPO$_4$, 14; KH$_2$PO$_4$, 6; MgCl$_2$, 0·2] containing glucose (0·25 % w/v) was used. The pH of the medium was kept constant using NaOH and it was acidified using HCl (1 M, Analar). Using this method we found that an abrupt drop in pH from 7·2 or 7·0 to 4·8

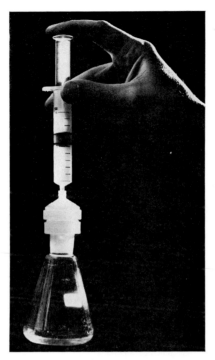

Fig. 9. The apparatus used for pH-shocking microbial cells.

did not bring about an adaptive response in either *E. coli* or *S. faecalis* (Fig. 8).

Continuous Culture

We have already pointed out that the use of batch culture for quantitative microbiology is subject to serious limitations, as it is not possible to specify and ensure constancy of important culture conditions. In the case of acid-producing microbes maintenance of a constant culture pH, even when buffers are used, is a major difficulty, as it cannot be achieved without altering the ionic composition of the medium. Continuous culture has been used as a means of obtaining steady-state growth under controlled conditions. For example Ellwood & Tempest (1972) found, using continuous culture, that culture pH value affected the cell wall composition of *Bacillus* spp. Medium is continuously added from a reservoir to a culture, contained in a growth vessel, maintained in a homogeneous state by stirring. We used a culture vessel with a working volume of 120 ml, kept constant by removing culture at the same rate that fresh medium was added. To ensure that the culture volume remained constant over a long period, a two channel peristaltic pump was used (LKB, Uppsala, Sweden), with one channel adding and the other channel removing medium from the vessel. As an additional safeguard a level control was fitted. We determined initially by observation whether the vessel progressively filled or emptied and to counter this a solenoid-operated pinch valve was fitted either to the medium inlet or exit. The valve could be set either normally open or normally closed—in this way the level was controlled to that set by the tip of the level electrode. It is an essential property of continuous cultures that for all dilution rates (*dil.*) which do not exceed the certain maximum growth rate (μ) value of the organism under those conditions the culture will adjust itself so that $\mu = dil.$ When $\mu = dil.$ the culture is in a steady state, when *dil.* exceeds μ wash-out will occur. The time needed for a culture to reach a steady state will be fixed by the dilution rate and the fractions of the original material remaining after 1, 2, 3 or 4 replacement times respectively are 0·367, 0·135, 0·050 and 0·015 (Pirt 1975). At a low dilution rate, of about 0·1, it will take 30 h for 95 % replacement of the medium. In those experiments where *dil.* was fixed we noticed that a small shift in pH (0·2 units) caused smaller oscillations of the culture density about its new value than when the pH was shifted by 2 units. At high dilution rates (about 1·0) the time for the cells to settle to a steady state may exceed the 95 % replacement time of the culture medium in the vessel.

Experimental details

The medium used in the continuous culture experiments was autoclaved in 12 l amounts (sufficient for 1000 h when *dil.* = 0·1 or 200 h when *dil.* = 0·5) to minimize any variations introduced by the medium itself. In experiments to investigate the effects of inhibitors we used an intermediate vessel (51) between the reservoir and the culture vessel.

All connections between the vessels were made using silicone tubing (Esco Ltd.) and all further connections made during culturing were steam sterilized. The apparatus, including the medium reservoir and pH electrode, was sterilized by autoclaving (45 min at 110°C). Glucose was added to the basal medium in the reservoir when both had cooled after autoclaving and the reservoir was chilled throughout the experiment by the circulation of chilled water through its heat exchangers. The pH in the working vessel was monitored using a Philips autoclavable pH electrode (CA42 HK-SK) and a Philips industrial non-indicating, Universal transmitter (PW 9406). Diagrams of the two types of continuous fermenter and their control mode are shown in Figs 10 & 13.

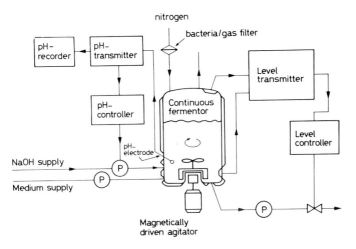

Fig. 10. A diagram of the apparatus used to grow *S. faecalis* at a constant growth
rate in order to determine the influence of pH on yield.

Yield

We grew *S. faecalis* at a range of pH values and fixed dilution rates (hence
fixed μ) using the equipment shown in Fig. 10. By determining molar
growth yield at fixed dilution rate and various pH values we estimated
the influence of pH on the cells' demand for maintenance energy (Fig.
11). It is only possible to do this in continuous culture when growth rate
(μ) can be fixed accurately over a range of pH values. Our previous work
using batch cultures has shown that both growth rate and yield were
altered by culture pH (Fig. 2). Forstel & Schleser (1976; Fig. 12) have
shown that the cells' demand for maintenance energy (shown by reduced
yield in our experiments) is increased at low growth rates, hence in batch
cultures where low pH reduces the growth rate it is impossible to
separate the effects of pH and μ on yield (Y). The pH in the fermentation
was controlled by the addition of sodium hydroxide (1 M), the production
of acid by the organisms was sufficient to lower the pH to the lowest
levels investigated (pH 4·2). The rate of alkali addition by an electrically
activated peristaltic pump (Watson-Marlow, Falmouth, England or
LKB, Uppsala, Sweden) was controlled using on/off control with a 30 s
integration time and the addition of 0·05 ml of alkali per pulse. Using
this method the pH was controlled to ±0·05 units. We estimated biomass
using a D: dry weight curve prepared previously, as the μ fixed experiments were done in similar broth to the batch culture experiments.

Based on the work of Rosenburger & Elsden (1960) we assumed that

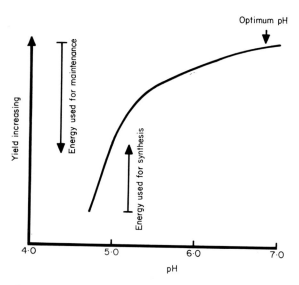

Fig. 11. The change in molar growth yield (glucose) at various pH values, determined using a continuous fermenter run at a fixed growth rate.

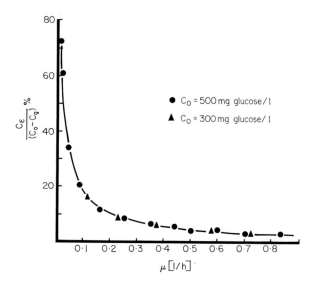

Fig. 12. Percentage substrate, $C\epsilon$, used in maintenance metabolism compared with the total consumption of the substrate (C_0-C_g) for *E. coli*, when C_0 = original substrate concentration and C_g = substrate remaining. (From Förstel & Schleser 1976.)

the ratio of opacity: dry weight for *S. faecalis* was unaffected by growth rate. Samples were collected and optical density measured as near to the sampling port as possible, as it was found that the *D* of the culture, after growth, changed during chilled storage. In the μ fixed experiments (initial level of glucose 5·5 mM) complete ($>98\%$) glucose utilization was found down to pH 4·8; below this pH value glucose ceased to be limiting.

Maximum growth rate (μ_{max})

The chemostat is not particularly suitable for growing microbes near to their maximum growth rates. At high dilution rates, approaching μ_{max}, the population is easily washed out (Martin & Hempfling 1976). We have used a method of continuous culture which employs growth-dependent pH changes to control the rate of addition of fresh medium to culture vessel (Fig. 13). The buffering capacity or pH of the incoming medium fixes the steady-state population density of the culture, but the growth rate ($=$dilution rate) is independent of the pH of the incoming medium. To justify the use of the pH auxostat which establishes μ_{max} by using acid production as the control mode, Martin & Hempfling (1976) derived an equation to link the rate of change of proton concentration in the medium with the rate of medium addition. They found reasonable agreement between the theory's predictions and their experimental results.

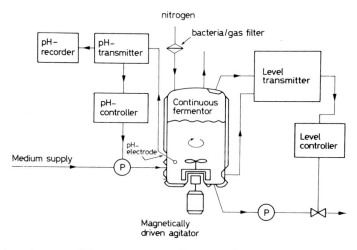

Fig. 13. A diagram of the apparatus used to grow *S. faecalis* at maximum growth rate (μ_{max}) in continuous culture at various pH values.

To ensure that only pH was growth-limiting we started with the original basal medium (see page 78) and increased the concentration of glucose, tryptone and yeast extract in it until further addition did not increase the growth rate at pH 5·5. The medium used in the μ_{max} experiments contained tryptone, 20 g l^{-1}; yeast extract 10 g l^{-1}; glucose 10 mg ml^{-1}. It was prepared and stored in the same way as the μ fixed medium and the pH corrected to 8·5 after autoclaving.

The rate of medium flow through the vessel was regulated according to the rate of deviation of culture pH from a fixed pH. This was monitored by a Philips autoclavable pH electrode (CA42 AK-SK) coupled to a Philips universal transmitter (PW 9406). The control signal was fed to a Philips Witronik PID (proportional integrating differentiating) controller which in turn controlled the rate of medium addition. The PID control was used to minimize displacement of, and oscillations about, the set point. The apparatus we used is shown diagrammatically in Fig. 13, typical results are shown in Fig. 14. The results of other workers are shown in Fig. 15. Using the equipment the accuracy of the pH control was ±0·1 pH units. The periodicity and amplitude of the oscillations increased immediately after the fixed point pH value was changed. As the steady state was approached the size of the oscillations in pH value about the fixed point diminished and their frequency decreased from one per hour to one per six hours. The rate of medium flow through the vessel oscillated slightly out of phase with the changes in pH, but the rate of glucose consumption exactly mirrored the changes in pH.

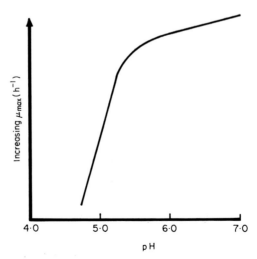

Fig. 14. The influence of pH on μ_{max} in steady-state continuous culture.

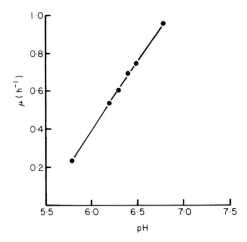

Fig. 15. The influence of pH on the steady-state maximum growth rate (μ_{max}) of a haemolytic streptococcus (Karush *et al.* 1956).

Maintenance energy

Accurate estimates of the influence of pH or any other stress factor on yield can only be made when growth rate is fixed. This is most easily done in continuous culture. In batch culture many factors may interact to influence yield. The use of yield as an index of microbial maintenance energy requirements assumes growth can be correlated closely not only with the amount of energy-yielding substrate consumed (Y), but also with the amount of ATP generated by its catabolism. The determination of molar growth yields has been used to provide an insight into metabolic differences between species and to allow assessment of the amount of useful energy (in ATP equivalents) generated from various substrates. We studied the yield of *S. faecalis* in order to determine the amount of energy employed for purposes other than biosynthesis when the organism is exposed to stress, in this case at high or low pH.

Under steady-state conditions the ratio of the specific growth rate (μ) and the specific rate of substrate consumption (γ) is fixed, however in a batch culture these two parameters may be uncoupled (Fig. 7). Experimental work using batch cultures has shown that whereas μ is dependent on γ the reverse is not true. Cells do not have to grow in order to consume substrate, in fact under some environmental conditions considerable amounts of substrate are consumed without apparent growth. Our work using batch and continuous culture (see also Forstel & Schleser 1976) has suggested that as growth rate is decreased, yield value progressively diminished. We have shown that not all the energy derived from catabolism was consumed in growth processes, but we have assumed that a small fraction—whose size was influenced by either μ or pH (separable in steady-state experiments) or both (shown in batch culture) was required for maintenance functions not directly associated with growth (see 'Uncoupling', Fig. 7). By subtracting the yield at any particular pH value from Y_{max} at optimum pH it is possible to estimate the maintenance energy requirement under those conditions. In this work we did not investigate the influence of μ or pH on changes in the physiological state in the cells, although without doubt these contribute to the change in yield. On the basis of this work it is not possible to say whether 'maintenance energy' is used 'vectorially' to pump protons from the cell or whether it is lost in slip reactions, because some factor other than the nutrients present in the medium is limiting growth. This work has shown that the proportion of the energy source removed from the medium by the cells and used for the synthesis of biomass is reduced by lowering either μ or pH. The methods used are simple, requiring a minimum of apparatus, and can provide an insight into the influence of stress on the economy of microbial cells.

Mutants

Owing to the ease with which microbial mutants can be obtained and the specific biochemical consequences of mutations, microbial mutants are now well established as tools for the elucidation of biochemical processes. Mutants allow the microbial physiologist to study particular biochemical processes whose importance cannot always be studied in isolation when metabolic inhibitors are used, as many inhibitors have unspecified secondary effects. We have investigated the influence of pH on growth using mutants with specific energy linked defects in their metabolism. We have also isolated pH-sensitive mutants.

Published mutants—nomenclature

Over the past decade studies on energetics including the biochemistry of energy-yielding metabolism, active transport and other membrane functions have led to the isolation of a range of *E. coli* mutants with defects in their energy-linked metabolism (Cox & Gibson 1974). Mutants in one such class are deficient in membrane-bound Mg^{++}-dependent ATPase activity. These mutants are characterized by their inability to couple respiration to ATP formation (i.e. oxidative phosphorylation) at the cytoplasmic membrane and may incidentally have impaired nutrient transport capabilities (Simoni & Postma 1975). Mg^{++}, dependent ATPase is the enzyme believed to catalyse reversible proton translocations across the cytoplasmic membrane, the direction of this membrane-bound ATPase activity being determined by a number of factors including the cells' internal pH and the environmental pH value. Energy-yielding hydrolysis of intracellular ATP may be used for biosynthesis or to expel protons from the cell; in reverse when protons enter the cell *via* the ATPase (because the cell interior is usually more alkaline than the environment) ATP may be formed or nutrients transported. The role of ATPase enzymes in microbial energetics has been thoroughly reviewed by Harold (1977).

Thus Mg^{++}-dependent ATPase deficient mutants have a reduced yield (Y) and lack one of the major mechanisms of proton transport across the cell membrane. They therefore represent ideal material for studying the role of ATPase linked processes and energy usage in general microbial growth at low pH.

Use of mutants to investigate the significance of cellular mechanisms in microbial resistance to low pH

We have used a Mg^{++}-dependent ATPase deficient mutant to investigate the effect of loss of ATPase activity (described above) on the ability of *E. coli* to grow at low pH. The wild type is termed ML308–225; its mutant, DL54, was first isolated and described by Simoni & Shallenberger (1972). DL54 lacks membrane bound Mg^{++}-dependent ATPase and is therefore believed to be defective in its ability to catalyse those transmembrane proton movements mediated by ATPase. Vesicles formed from the membrane of this mutant are also abnormally permeable to protons (Altendorf *et al.* 1974).

We grew the wild type and mutant in a minimal broth medium $(g\ l^{-1})$: $(NH_4)_2SO_4$, $2\ g\ l^{-1}$; K_2HPO_4, $14\ g\ l^{-1}$; KH_2PO_4, $6\ g\ l^{-1}$; $MgCl_2$, $0.2\ g\ l^{-1}$; glucose 0.25% (w/v). The overnight culture was depleted of endogenous energy reserves by the method of Wilson *et al.* (1976) with the following modifications. The minimal broth described above was used in place of medium 63 (Wilson *et al.* 1976), washed cells were suspended at a density of $D = 0.5$ (at 580 nm), and depleted cells were washed twice and resuspended in a buffer solution containing K_2HPO_4 (100 mM) and NaH_2PO_4 (100 mM) adjusted to pH 7.0 to give a final optical density 0.5.

Cell suspension (0.02 ml) was used to inoculate: (1) the minimal medium described above containing 1.5% Noble Agar (Difco); (2) Heart Infusion Agar (Difco), both adjusted to a range of pH values by the addition of HCl. The lowest pH allowing growth was determined by plate assay after 48 h incubation at 37°C. There was only 0·1 pH unit difference between the lower growth limiting pH of the wild type and its ATPase deficient mutant (Table 1) indicating that in *E. coli* the presence of a membrane bound Mg^{++}-dependent ATPase is not essential for

TABLE 1

Lowest pH-values for growth of a wild-type Escherichia coli (*ML 308-225*) *and its ATPase-deficient mutant (DL 54)*

Medium	Lowest pH for growth of	
	Strain ML308-225	Strain DL54
Minimal agar medium	4·5	4·6
Heart infusion agar	4·4	4·5

aerobic growth at low pH. However DL54 grew very poorly under aerobic conditions. Similar results have been obtained by other workers using mutants lacking Mg^{++}-activated ATPase (Yamamoto *et al.* 1973). The results illustrate the importance of respiration as a means of energizing the cell membrane for nutrient transport and proton expulsion in the absence of membrane bound ATPase activity.

After each experiment, the mutant culture DL54 was checked for the presence of revertant cells by plating onto minimal agar containing sodium malate (0·5 % w/v) and sodium succinate (0·5 % w/v) as the sole carbon sources. ATPase mutants are unable to grow on these carbon sources (which must be used oxidatively). The wild-type, and any revertants of DL54, will be able to grow on such medium.

Isolation procedure

We have isolated mutants of *E. coli* NCIB 10430 $T^-A^-U^-$ unable to grow at low pH. This strain requires the addition of thymine (T), arginine (A) and uracil (U) (25 μg ml^{-1}) to minimal medium to ensure adequate growth. The isolation procedure we used is outlined in Fig. 16. *Escherichia coli* 10430 and K12 were grown to mid-logarithmic phase in minimal broth (described previously) at pH 7 and 37°C then mutagen treated with either ethyl methanesulphonate (EMS, Koch-Light Laboratories Ltd., Colnbrook, SL3 OB2, UK) or *N*-methyl-*N*-nitroso-*N*-nitroguanidine (NTG, Koch-Light) as shown in Fig. 16. The cultures were then washed twice with distilled water to free them of mutagen and incubated overnight in minimal broth at pH 7. This incubation step allows for the segregation of mutant cells into pure clones. See Hopwood (1970) for a review of mutant isolation techniques.

Method

Penicillin Lethal Enrichment of <u>Escherichia coli</u>
K12 and 15 T⁻A⁻U⁻

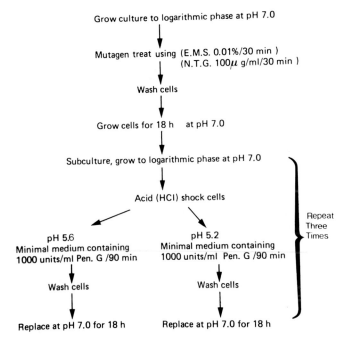

Fig. 16. The procedure used to select for pH sensitive mutants of *E. coli*.

After overnight growth the cell suspension was subcultured into minimal broth (pH 7), grown to the mid-logarithmic stage of growth and then
acid shocked, using the method described above, to the values indicated.
Penicillin was added to these cultures as indicated in Fig. 16. After
90 min the cultures were washed twice in distilled water to remove the
penicillin prior to reincubating the cells in minimal broth (pH 7) overnight. This penicillin lethal-enrichment technique was used to isolate
mutants sensitive to low pH. Penicillin will only kill growing cells, hence
in this culture system only those cells whose growth was inhibited by
the low pH values should survive. The enrichment step was repeated
three times to increase the number of acid-sensitive cells relative to wild
type cells sufficiently to allow selection of mutants. Cultures were then
diluted serially with distilled water and 0·1 ml of the dilute suspension

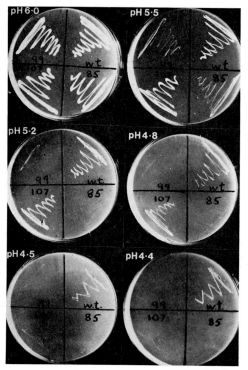

Fig. 17. The sensitivity of mutants of *E. coli* (NCIB 10430) T⁻ A⁻ U⁻ (see p. 96) to low pH. The mutants 85, 99 and 107 were obtained using the selection procedure shown in Fig. 16.

was spread onto the surface of a minimal agar plate (pH 7). After 2 days incubation at 37°C those plates containing 100–300 colonies were replica plated onto the same medium adjusted to low pH values. Those colonies unable to grow at low pH were isolated for further examination.

We used a flat bottomed surface combination pH electrode (Gorski & Ritzert 1977; manufactured by Owens-Illinois Inc, Toledo, Ohio-43666, USA) for determining the pH of agar plates prior to replica plating.

We obtained acid sensitive mutants of *E. coli* NCIB 10430 T⁻A⁻U⁻ showing a range of sensitivity to acid (HCl) pH. Figure 17 illustrates the sensitivity of three mutants (85, 107 and 99) compared to the wild type (wt) parent culture. We are using such mutants to investigate the biochemical processes required for growth at low pH.

We thank Dr R. D. Simoni, Stanford University, California, USA for providing ML308-225 and its mutant derivative DL54.

References
ALTENDORF, K., HAROLD, F. M. & SIMONI, R. D. 1974 Impairment and restoration of the energised state in membrane vesicles of a mutant of *Escherichia coli* lacking adenosine triphosphatase. *Journal of Biological Chemistry* **249**, 4587–4593.

BAUCHOP, T. & ELSDEN, S. R. 1960 The growth of micro-organisms in relation to their energy supply. *Journal of General Microbiology* **23**, 457–469.

BROCK, T. D. 1969 Microbial growth under extreme conditions. *Symposium of the Society of General Microbiology* **19**, 15–41.

COX, G. B. & GIBSON, F. 1974 Studies on electron transport and energy-linked reactions using mutants of *Escherichia coli*. *Biochimica et Biophysica Acta* **346**, 1–25.

ELLWOOD, D. C. & TEMPEST, D. W. 1972 Influence of culture pH on the content and composition of teichoic acids in the walls of *Bacillus subtilis*. *Journal of General Microbiology* **73**, 395–401.

FÖRSTEL, H. & SCHLESER, G. 1976 Role of maintenance metabolism for biomass production. *Abstracts of the Fifth International Fermentation Symposium, Berlin*.

GARLAND, P. B. 1977 Energy transduction and transmission in microbial systems. *Symposium of the Society for General Microbiology* **27**, 1–21.

GOMEZ-PUYOU, A. & GOMEZ-LOJERO, C. 1977 The use of ionophores and channel-formers in the study of the function of biological membranes. *Current Topics in Bioenergetics* **6**, 221–257.

GOOD, N. E., WINGET, G. D., WINTER, W., CONNOLLY, T. N., IZAWA, S. & SINGH, R. H. M. 1966 Hydrogen-ion buffers for biological research. *Biochemistry* **5**, 467–477.

GORSKI, T. W. & RITZERT, R. W. 1977 pH Measurements of agar culture media using a flat-bottom electrode. *Applied and Environmental Microbiology* **34**, 242–243.

GOULD, G. W. & MEASURES, J. C. 1977 Water relations in single cells. *Philosophical Transactions of the Royal Society of London Series B* **278**, 151–166.

HAMILTON, W. A. 1975 Energy coupling in microbial transport. *Advances in Microbial Physiology* **12**, 1–53.

HAROLD, F. M. 1977 Membranes and energy transduction in bacteria. *Current Reviews in Bioenergetics* **6**, 83–149.

HAROLD, F. M. & BAARDA, J. R. 1968 Inhibition of membrane transport in *Streptococcus faecalis* by uncouplers of oxidative phosphorylation and its relation to proton conduction. *Journal of Bacteriology* **96**, 2025–2034.

HAROLD, F. M., BAARDA, J. R. BARON, C. & ABRAMS, A. 1969 Dio-9 and chlorhexidine: Inhibitors of membrane-bound ATPase and of cation transport in *Streptococcus faecalis*. *Biochimica et Biophysica Acta* **183**, 129–136.

HOPWOOD, D. A. 1970 The isolation of mutants. In *Methods in Microbiology* Vol. 3A eds Norris, J. R. & Ribbons, D. W. pp. 363–434, London: Academic Press.

KARUSH, F., IACOCCA, V. F. & HARRIS, T. N. 1956 Growth of group A haemolytic *Streptococcus* in the steady state. *Journal of Bacteriology* **72**, 283–294.

LANGWORTHY, T. A. 1978 Microbial life at extreme pH values. In *Microbial Life in Extreme Environments* ed. Kushner, D. J. Ch. 7, London: Academic Press.

MALLETTE, M. F. 1970 Evaluation of Growth by Physical and Chemical Means. In *Methods in Microbiology* Vol. 2, eds Norris, J. R. & Ribbons, D. W., London: Academic Press.

MARTIN, G. A. & HEMPFLING, W. P. 1976 A method for the regulation of microbial population density during continuous culture at high growth rates. *Archives of Microbiology* **107**, 41–47.

MOSSEL, D. A. A. 1971 Physiological and metabolic attributes of microbial groups associated with foods. *Journal of Applied Bacteriology* **34**, 95–118.

MOUSTAFA, H. M. & COLLINS, E. B. 1968 Molar growth yields of certain lactic acid bacteria as influenced by autolysis. *Journal of Bacteriology* **96**, 117–125.

PIRT, S. J. 1975 Energy and carbon source requirements. In *Principles of Microbe and Cell Cultivation*. Ch. 8, Oxford: Blackwell Scientific Publications.

ROSENBERGER, R. F. & ELSDEN, S. R. 1960 The yields of *Streptococcus faecalis* grown in continuous culture. *Journal of General Microbiology* **22**, 726–739.

SIMONI, R. D. & SHALLENBERGER, M. K. 1972 Coupling of energy to active transport of amino acids in *Escherichia coli*. *Proceedings of the National Academy of Sciences U.S.A.* **69**, 2663–2667.

SIMONI, R. D. & POSTMA, P. W. 1975 The energetics of bacterial active transport. *Annual Review of Biochemistry* **44**, 523–554.

STOUTHAMER, A. H. 1970 Determination and significance of molar growth yields. In *Methods in Microbiology* Vol. 1, eds Norris, J. R. & Ribbons, D. W., London: Academic Press.

STOUTHAMER, A. H. & BETTENHAUSSEN, C. 1973 Utilisation of energy for growth and maintenance in continuous and batch cultures of micro-organisms. *Biochimica et Biophysica Acta* **301**, 53–70.

TEMPEST, D. W. 1978 The biochemical significance of microbial growth yields: a reassessment. *Trends in Biochemical Sciences*, August 1978, 180–184.

WILSON, D. M., ALDERETE, J. F., MALONEY, P. C. & HASTINGS WILSON, T. 1976 Protonmotive force as the source of energy for adenosine 5′-triphosphate synthesis in *Escherichia coli*. *Journal of Bacteriology* **126**, 327–337.

YAMAMOTO, T. H., MEVEL-NINIO, M. & VALENTINE, R. C. 1973 Essential role of membrane ATPase or coupling factor for anaerobic growth and anaerobic active transport in *Escherichia coli*. *Biochimica et Biophysica Acta* **314**, 267–275.

Growth in Hyperbaric Oxygen: Effect upon the Resistance to Antibacterial Agents and Morphology of *Pseudomonas aeruginosa*

M. A. Kenward* and S. R. Alcock

Department of Bacteriology, The Medical School, Foresterhill, Aberdeen, UK

M. R. W. Brown

Department of Pharmacy, University of Aston in Birmingham, Gosta Green, Birmingham, UK

The systems used in high-pressure microbiological research can be broadly classed into those concerned with hydrostatic pressure and those concerned with hyperbaric pressure. Interest in the effect of hydrostatic pressure upon bacteria arose mainly from the field of deep sea marine biology and many systems were developed to enable sampling and study of bacterial populations from deep sea environments.

Hydrostatic Systems

Essentially hydrostatic systems allow the placing of a bacterial suspension in a cylinder which is closed at one or both ends by a moveable piston. The external pressure on the piston is increased either by direct mechanical force or by immersion in a fluid-filled pressure vessel which is in turn pressurized.

The simplest systems consist of stoppered glass tubes or sealed syringes (Yayanos 1969; Schwartz *et al.* 1975; Albright 1975; Arcuri & Ehrlich 1977) which are placed in a fluid or gas-filled pressure vessel. Most of these types seem to be based on the apparatus described by Zobell & Oppenheimer (1950). More sophisticated purpose-built systems that allow *in situ* monitoring and non-decompressive sampling have been reported by Taylor & Jannasch (1976) and Jannasch *et al.* (1976).

* Present address: Department of Biological Sciences, Wolverhampton Polytechnic, Wolverhampton, UK.

The reviews of Zimmerman (1971) Morita (1970; 1976) and Marquis (1976) provide further details and comprehensive references about these systems.

Hyperbaric Systems

Apparatus to study the effect of hyperbaric pressure basically consists of chambers into which cultures in liquid or on solid media are placed. The chamber is pressurized by introducing a gas phase until the required pressure is reached. Hyperbaric chambers are used both as 'closed' systems or as 'open' systems. With the latter the pressure is maintained by an exhaust valve which allows a flow of fresh gas through the chamber. Hyperbaric chambers of open and closed type can be readily adapted from existing laboratory equipment such as anaerobic jars (Kaye 1967), autoclaves (McAllister *et al.* 1963; Bornside 1967) or ethylene oxide sterilizers (Pakman 1971). More sophisticated chambers can be obtained commercially (Brown *et al.* 1977; Harley *et al.* 1978) or built from stainless steel (Hopkinson & Towers 1963) or "Plexiglass" (methacrylates; Gordon & Gillmore 1974).

Most of the studies on elevated partial pressures of oxygen (pO_2) have made use of hyperbaric chambers in which the pO_2 can be controlled by the manipulation of the gas mixture (Gottlieb 1971). An advantage of hyperbaric chambers is that bacterial growth can be followed for extended periods without oxygen limitation becoming a problem. However O_2 limitation can be overcome in hydrostatic systems by the use of oxygen-saturated fluoro-carbon liquids which act as an oxygen reservoir and a "sink" for metabolic gases (see Marquis 1976 for further references).

We have used a simple static hyperbaric chamber to study the effect of hyperbaric oxygen upon growth of *Pseudomonas aeruginosa* and its drug resistance. A summary of our results follows.

Hyperbaric Oxygen Effects on *Pseudomonas aeruginosa*
(Kenward *et al.* 1978)

Incubation of *P. aeruginosa* on blood agar plates in 2 atm absolute (ATA) of 100% oxygen at 37°C for 24 h produced colonies of abnormal morphology. Plates inoculated with about 10^5 cells gave rise to small atypical colonies ('dwarf'). Inocula of about 10^9 cells produced about 100 large, domed, mucoid colonies ('giant') growing on a lawn of dwarfs. Giants, but not dwarfs, showed pigment production. Neither variant showed β-haemolysis of the blood after 24 h incubation in hyperbaric oxygen (HBO), but the giants showed haemolysis after 48 h in HBO. Control plates incubated for the same period in air at ambient pressure showed normal colony or lawn formation, extensive haemolysis and pyocyanin production.

Suspensions of cells prepared from dwarf colonies were more resistant than suspensions from normal colonies when plated on to agar containing polymyxin, tetracycline or phenoxyethanol and incubated in air. Suspensions of cells prepared from giant colonies showed a response to these antimicrobial agents that was between that of dwarf and normal suspensions.

Pre-incubation of similarly inoculated antibiotic plates for 24 h in HBO resulted in an enhancement of the action of polymyxin and phenoxyethanol and, to a lesser extent, of tetracycline.

The resistance of cells from variant colonies to the drugs was lost after one subculture in broth or on agar in the absence of the drug. The variant colonies did not breed true in air or HBO suggesting that the effect of HBO upon *P. aeruginosa* was phenotypic rather than genotypic.

References

ARCURI, E. J. & EHRLICH, H. L. 1977 Influence of hydrostatic pressure on the effects of heavy metal cations of manganese, copper, cobalt and nickel on the growth of three deep-sea bacterial isolates. *Applied and Environmental Microbiology* **33**, 282–288.

ALBRIGHT, L. J. 1975 The influence of hydrostatic pressure upon biochemical activities of heterotrophic bacteria. *Canadian Journal of Microbiology* **21**, 1406–1412.

BORNSIDE, H. 1967 Exposure of *Pseudomonas aeruginosa* to hyperbaric oxygen: inhibited growth and enhanced activity to polymyxin B. *Proceedings of the Society of Experimental Biology and Medicine* **125**, 1152–1156.

BROWN, O. R., YEIN, F., MATHIS, R. & VINCENT, K. 1977 Oxygen toxicity: comparative sensitivities of membrane transport bioenergetics and synthesis in *Escherichia coli*. *Microbios* **18**, 7–25.

GORDON, F. B. & GILLMORE, J. D. 1974 Parabarosis and experimental infections. 4. Effect of varying O_2 tensions on Chlamydial infection in mice and cell cultures. *Aerospace Medicine* **45**, 257–262.

GOTTLIEB, S. F. 1971 Effect of hyperbaric oxygen on micro-organisms. *Annual Reviews of Microbiology* **25**, 111–152.

HARLEY, J. B., SANTANGELO, G. M., RASMUSSEN, H. & GOLDFINE, H. 1978 Dependence of *Escherichia coli* hyperbaric oxygen toxicity on the lipid acyl chain composition. *Journal of Bacteriology* **134**, 808–820.

HOPKINSON, W. I. & TOWERS, A. G. 1963 Effect of hyperbaric oxygen on some common pathogenic bacteria. *The Lancet II*, 1361–1363.

JANNASCH, H. W., WIRSEN, C. O. & TAYLOR, C. D. 1976 Undecompressed microbial populations from the deep sea. *Applied and Environmental Microbiology* **32**, 360–367.

KAYE, D. 1967 Effect of hyperbaric oxygen on aerobic bacteria *in vitro* and *in vivo*. *Proceedings of the Society for Experimental Biology and Medicine* **124**, 1090–1093.

KENWARD, M. A., ALCOCK, S. R. & BROWN, M. R. W. 1978 Effects of hyperbaric oxygen upon colonial morphology and resistance of *Pseudomonas aeruginosa*. *Proceedings of the Society for General Microbiology* **5,** 55.

MARQUIS, R. E. 1976 High-pressure microbial physiology. *Advances in Microbial Physiology* **14,** 159–241.

MCALLISTER, T. A., STARK, J. M., NORMAN, J. N. & ROSS, R. M. 1963 Inhibitory effects of hyperbaric oxygen on bacteria and fungi. *The Lancet II,* 1040–1042.

MORITA, R. Y. 1970 In *Methods in Microbiology*. Vol. 2 eds Norris, J. R. & Ribbons, D. W. pp. 243–257, London: Academic Press.

MORITA, R. Y. 1976 Survival of bacteria in cold and moderate hydrostatic pressure environments with special reference to psychrophilic and barophilic bacteria. *Symposium of the Society for General Microbiology* **26,** 279–298.

PAKMAN, L. M. 1971 Inhibition of *Pseudomonas aeruginosa* by hyperbaric oxygen. I. Sulphonamide activity enhancement and reversal. *Infection and Immunity* **4,** 479–487.

SCHWARZ, J. R., WALKER, J. D. & COLWELL, R. R. 1975 Deep sea bacteria: growth and utilization of n-hexadecane at *in situ* temperature and pressure. *Canadian Journal of Microbiology* **21,** 682–687.

TAYLOR, C. D. & JANNASCH, H. W. 1976 Subsampling technique for measuring growth of bacterial cultures under high hydrostatic pressure. *Applied and Environmental Microbiology* **32,** 355–359.

YAYANOS, A. A. 1969 A technique for studying biological reaction rates at high pressure. *Reviews of Scientific Instrumentation* **40,** 961–963.

ZIMMERMAN, A. M. 1971 High-pressure studies in cell biology. *International Reviews of Cytology* **30,** 1–47.

ZOBELL, C. E. & OPPENHEIMER, C. H. 1950 Some effects of hydrostatic pressure on the multiplication and morphology of marine bacteria. *Journal of Bacteriology* **60,** 771–781.

Xerotolerant Yeasts at High Sugar Concentrations

R. H. Tilbury

Tate & Lyle Ltd., Group Research and Development, Reading, Berkshire, UK

It is generally accepted that water activity (a_w), not osmotic pressure, is the principal factor governing the water relations of micro-organisms (Mossel 1975). Yeasts capable of growth at low a_w values or high solute concentrations have been classified variously as osmophilic (Christian 1963), osmotophilic (van der Walt 1970), osmotolerant (Anand & Brown 1968), osmoduric (van der Walt 1970) osmotrophic (Sand 1973), xerophilic (Pitt 1975) and xerotolerant (Brown 1976). Strictly the term 'xerotolerant' should be used in preference to 'osmophilic' and other terms because these yeasts do not have a general requirement for decreased a_w, but merely tolerate drier conditions better than do non-osmotolerant yeasts (Anand & Brown 1968). However, the name osmophilic is most commonly used and will be adopted here for convenience.

Various definitions of osmophilic yeast have been proposed, usually based on an organism's ability to grow at a particular a_w value or sugar concentration. Christian (1963) defined osmophilic yeasts as those capable of multiplication in concentrated syrups of a_w below 0·85 whilst Scarr & Rose (1966) chose a limit of 65° Brix sucrose, equivalent to an a_w of 0·865 at 25°. An a_w of 0·87 was regarded by Mossel (1971) as the lowest limit for growth of non-osmophilic yeasts. Van der Walt (1970) recommended the use of agar media containing 50% (w/w) and 60% (w/w) glucose to distinguish between 'osmotolerant' and 'osmotophilic' yeasts, equivalent to a_w values of approximately 0·90 and 0·85 respectively. In the sugar industry, osmophilic yeasts are conveniently defined in practical terms, as those capable of growth in a saturated sucrose solution. This is approximately 67°C Brix at 25°C, equivalent to an a_w of 0·85. Finally, Pitt (1975) defined xerophilic fungi, including osmophilic yeasts, as those 'capable of growth, under at least one set of environment conditions, at an a_w below 0·85'. Choice of this limit was empirical, but in view of its practical value, it will be used here.

Since the classical paper by Scott (1957), there have been published several reviews dealing with osmophilic yeasts and/or their role in food spoilage, notably those of Ingram (1957; 1958); Christian (1963); Onishi (1963); Walker & Ayres (1970), Kushner (1971) and Tilbury (1976). The Ph.D. theses of Anand (1969), Koh (1972) and Corry (1974) contain useful critical reviews, mainly concerned with physiological aspects of osmophilic yeasts. Spoilage of foods of plant origin by xerophilic fungi was reviewed by Pitt (1975), and Mossel (1975) discussed the relationships between water and micro-organisms in foods.

A discussion of the mechanisms of resistance of xerotolerant yeasts to water stress is out of place here, but this has been comprehensively reviewed by Brown (1976).

Water Activity

It is helpful to understand the concept of water activity (a_w) and be able to measure or estimate the a_w of a commodity, in order to assess its susceptibility to spoilage by osmophilic yeasts. Useful references are the papers by Scott (1957), Ayerst (1965), Bone (1973), Mossel (1975), Ross (1975), Hardman (1976) and Brown (1976).

Briefly $a_w = P_w/P_w{}^o$ and, in a sufficiently dilute solution, where Raoult's law holds:

$$\frac{P_w}{P_w{}^o} = \frac{n_1}{n_1 + n_2}$$

where P_w = vapour pressure of water over a solution; $P_w{}^o$ = vapour pressure of pure water at the same temperature; n_1 = no. of moles of solvent; n_2 = no. of moles of solute.

Water activity is numerically equal to the equilibrium relative humidity (ERH, per cent) above a material in a confined space, expressed as a decimal. It is related to osmotic pressure (π) by the equation:

$$\pi = -\frac{RT \log_e a_w}{\overline{V}}$$

where R = gas constant per mole; T = absolute temperature; $\overline{V}$ = partial molal volume of water.

In complex solutions, a_w may be estimated by the equation:

$$\log_e a_w = -\frac{X_1 M_1 \phi_1 + X_2 M_2 \phi_2 + \text{etc.}}{55 \cdot 51}$$

where X_1, X_2 = no. of ions generated by solutes 1 and 2 where the solutes dissociate in solution; M_1, M_2 = molality of solutes 1 and 2; ϕ_1, ϕ_2 = molal osmotic coefficients of solutes 1 and 2, from tables (Robinson & Stokes 1965); $55 \cdot 51$ = molality of water (solvent).

Alternatively, using the equation of Ross (1975):

$$a_w = (a_w{}^o, \text{ solute 1}) \times (a_w{}^o, \text{ solute 2}) \times \text{etc.}$$

These equations are useful for estimating the a_w of media in growth experiments with osmophilic yeasts. Data on the relationship between concentration and a_w for various sugars and glucose syrups was presented by Norrish (1966) and W. J. Kooiman (pers. comm., 1978) and is summarized in Fig. 1. Note that concentration can be expressed as % (w/v), % (w/w) (Brix), molarity or molality; care must be taken to differentiate between them.

Estimates of a_w of materials should always be confirmed experimentally because many solutes deviate from ideality at high concentrations. Classical methods of determining a_w (ERH), such as those based on dew point measurement (Anagnostopoulos 1973), or 'sorption rate' (Landrock & Proctor 1951), suffer from various disadvantages, discussed by Ayerst (1965) and Hardman (1976). Instrumental methods are now available which can overcome most of these disadvantages, reviewed by Hardman (1976). In our laboratory an electrolytic hygrometer, the SINA Equi-Hygroscope (Nova-Sina, Zürich), is successfully used for the routine determination of a_w of sugars, syrups and related products (Troller 1977).

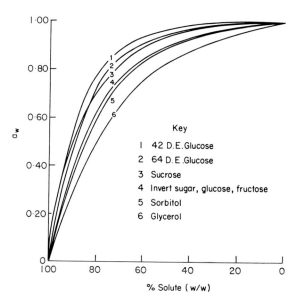

Fig. 1. Relationship between solute concentration and water activity of some sugars, glucose syrups and polyols (after Norrish 1966). Reproduced by kind permission of the *Journal of Food Technology*.

Principal Species of Osmophilic Yeast

Table 1 lists those species of osmophilic yeast which are most commonly associated with spoilage of high sugar content foods. The data are partly extracted from the review by Walker & Ayres (1970) and partly from additional or more recent publications (Tilbury 1976). The yeasts are named and classified according to Lodder (1970), but common synonyms and names of perfect/imperfect forms are included for ease of reference to earlier literature.

Some attempt has been made to indicate the sugar and salt (NaCl) tolerance of the organisms, using maximum concentrations cited for the species in Lodder (1970). This information is incomplete as in some genera only salt tolerance is quoted, in others only sugar tolerance, and in several genera combined salt and sugar tolerance is tested. In Lodder, sugar tolerance is given in terms of ability to grow in media containing 50% and 60% (w/w) glucose, equivalent to a_w values of about 0·90 and 0·845 respectively; these values are not necessarily the maximum sugar concentration tolerated. For comparison, a_w values of salt concentrations are as follows: 10% (w/v) $= a_w$ 0·94; 15% $= 0·895$ and 20% $= 0·845$. Additional information on sugar tolerance is taken from three sources: Scarr & Rose (1966) identified yeasts capable of growth in 65% (w/w) sucrose syrups; Tilbury (1967) determined the minimum a_w for growth in sucrose/glycerol syrups and Windisch & Neumann (1965) tested isolates for ability to ferment fructose solutions containing 45, 60 and 75% sugar. According to the terminology of van der Walt (1970) yeasts that can grow at concentrations of 50% but not 60% (w/w) glucose are called osmoduric or osmotolerant, whereas those capable of growth at 60% (w/w) glucose, are osmophilic. This principle was adopted by Davenport (1975; pers. comm. 1976) but the names were shortened to 'osmotolerant' and 'osmophilic'. On this basis, only some of the species listed in Table 1 may be classified as truly osmophilic. According to Davenport (pers. comm. 1976) the principal osmophilic yeasts are as follows: S. rouxii, S. bailii var. osmophilus; S. bisporus var. mellis; T. lactis-condensi; Schizosacch. pombe; D. hansenii; T. candida; Pi. ohmeri and H. anomala var. anomala. Many species can tolerate high concentrations of both sugar and salt.

Some important physiological properties of the main osmophilic yeasts are summarized in Table 2. This information is useful in assessing the spoilage potential of particular foods by osmophilic yeasts.

TABLE 1

Osmotolerant and osmophilic yeasts, according to the classification and descriptions of Lodder (1970)

| Organism | Synonyms; perfect/ imperfect forms | a_w tolerance (max. solute conc^n or min. a_w for growth) | | Type of commodity commonly spoilt | |
		Glucose (% w/w)	NaCl (% w/w)	High sugar	High salt
Candida guilliermondii	(P) *Pi. guilliermondii*	57[a]	13	+	
C. krusei			10	+	+
C. parapsilosis		60	17	+	
C. tropicalis			13	+	
C. valida	(P) *Pi. membranefaciens*		9	+	+
Citeromyces matritensis	(I) *T. globosa*	57;[a]	(5 + 10)	+	
Debaryomyces hansenii	(I) *T. candida*	50		+	+
Endomycopsis burtonii	*E. chodatii*	50		+	
Hansenula anomala var. anomala	(I) *C. pelliculosa*	0·75[b]	(5 + 10)	+	+
H. polymorpha			(5 + 10)		
H. subpelliculosa			(5 + 10)	+	+
Kloeckera apiculata		50		+	
Pichia farinosa		50		+	
Pi. guilliermondii	(I) *C. guilliermondii*	50		+	
Pi. membranefaciens	(I) *C. valida*	50		+	+
Pi. ohmeri	(I) *E. ohmeri*	50		+	+
Saccharomyces bailii var. bailii	*S. acidifaciens*; *S. elegans*	50		+	+
S. bailii var. osmophilus		60		+	+
S. bisporus var. bisporus		50		+	
S. bisporus var. mellis	*S. mellis*	0·70,[b] 60		+	
S. microellipsoides var. osmophilus		60		+	
S. rosei		60		+	+
S. rouxii	many *Zygosacch.* spp.	0·65, 60		+	+
Schizosaccharomyces octosporus		50		+	
Schizosacch. pombe		50		+	
Torulopsis apicola	*T. bacillaris*	57[a]	11	+	
T. candida	(P) *D. hansenii* *T. famata*	0·65[b]	21	+	+
T. colliculosa		75	13	+	
T. dattila			15	+	
T. etchellsii		0·70[b]	21	+	+
T. glabrata			13	+	
T. kestoni		57[a]		+	
T. lactis-condensi		57[a]	11	+	
T. magnoliae			11	+	
T. versatilis	*Brett. versatilis*	0·70[b]	13	+	+

(I), Imperfect form; (P), Perfect form.

[a] 57% (w/w) glucose = 65° Brix sucrose = a_w 0·865. Data taken from Scarr & Rose (1966).

[b] Minimum a_w for growth in sucrose/glycerol syrups, determined by Tilbury (1967).

TABLE 2

Some physiological properties of common osmophilic yeasts (Lodder 1970)

Physiological characteristic		*S. bailii* var. *osmophilus*	*S. bisporus* var. *mellis*	*S. rouxii*	*D. hansenii* and *T. candida*	*H. anomala* var. *anomala*
Sugar fermenta-tion and assimila-tion	glucose	F	F	F	FVW; A	F
	galactose	AV	AV	A	FVW; A	F
	sucrose	FV; AV	FV; AV;W	FV; AV	FVW; A	F
	maltose	NA	NA	F	FVW; A	FV; A
	lactose	NA	NA	NA	AV	NA
	raffinose	FV; AV	NA	NA	A	F
Carbon assimila-tion	soluble starch	NA	NA	NA	AV	A
	ethanol	A	A	AV	A	A
	glycerol	A	A	AV	A	A
	sorbitol	A	A	A	A	A
Assimilation of KNO_3		NA	NA	NA	NA	A
Growth in vitamin-free medium		—	—	—	V	+
Growth at 37°C		—	—	—	V	V

F, fermented and assimilated; NA, not assimilated; A, assimilated but not fermented; W, weak; V, variable: some strains positive, others negative.

Methodology

Media and methods for isolation and enumeration of osmophilic yeasts

There are no commercially-available dehydrated selective media for osmophilic yeasts, but a number of suitable media have been described in the literature. The important features of medium composition, as discussed by Sand (1973) are its a_w, pH, redox potential and nutrient availability. Obviously a_w is the key factor which enables selection of osmophilic yeasts and inhibition of bacteria, but a relatively low pH is also useful. Ingram (1959) described a 50% glucose, citric acid, tryptone agar for isolation of osmotolerant yeasts from concentrated orange juice, but it proved tedious to prepare. In our laboratory we routinely use Scarr's osmophilic agar for enumeration of osmophilic yeasts in sugar products (Scarr 1959). It is simple to prepare and easy to use, but due to its relatively high a_w of *ca*. 0·95 it occasionally permits bacterial growth; certainly it will support growth of some non-osmophilic yeasts. Van der Walt (1970) recommends 50% and 60% (w/w) glucose agars, at a_w values of *ca*. 0·90 and 0·845 for characterization of osmotolerant and osmophilic yeasts respectively. Mossel (1951) prefers a 60% (w/w) fructose peptone meat-extract agar, at an a_w of 0·84 since fructose is more soluble than glucose. Most recently Pitt (1975) described a number of media suitable for cultivation of xerophilic fungi; one was a malt extract yeast extract agar containing 60% (w/w) glucose, a_w 0·845.

Scarr's osmophilic agar is prepared as follows: Make up a 45° Brix syrup containing sucrose 35 parts, glucose 10 parts and water 55 parts by weight. Dissolve Oxoid Wort Agar (CM 247) in this solution (50 g l^{-1} syrup) and sterilize by autoclaving at 115°C for 15 min. Reinforcement of gel strength by addition of plain agar (10 g l^{-1}) is desirable where plates are to be used for streaking or spreading. The pH of Scarr's agar is usually *ca.* 4·8.

Plates are incubated at 25–30°C and counted after 3 and 5 days. Colonies on this medium are usually well defined and opaque. Osmophilic *Saccharomyces* species form colonies 1–2 mm in diam. after 3 days, whilst the slower growing *Torulopsis* spp. require 5 days. Examples of colony morphology are shown in Fig. 2.

For routine enumeration of true osmotolerant yeasts in products, Scarr's agar has the convenience of giving a result within 3–5 days, but has the disadvantage of lack of specificity. On the other hand, media of a_w 0·85 possess the advantage of specificity but the disadvantage that incubation periods of up to 28 days are needed. In our laboratory we have attempted to formulate a medium which will select only true osmotolerant yeasts within a period of 3 days incubation at 30°C. Representative species of true osmotolerant yeasts (6 isolates) and yeasts which could grow on Scarr's agar but not in 65° Brix sucrose syrup (12 isolates) were selected from the Tate & Lyle culture collection. Cultures were streaked on Scarr's agar plates modified by increasing the sucrose content in units of 5° Brix from 45 to 65° Brix. The preliminary results (unpublished) showed that at 50° Brix all the true osmotolerant yeasts exhibited good growth within 3 days whilst none of the non-osmotolerant yeasts had grown. Confirmatory work needs to be done on this modification of Scarr's agar.

Viable counts of osmotolerant yeasts may be done by traditional techniques such as membrane filtration, direct plating, or by serial dilution followed by pour plates or spread plates. Techniques suitable for use with fruit juice concentrates were described by Sand (1973) but they are of general applicability. Two special features should be noted. Firstly, in order to avoid loss of viability of cells due to osmotic shock in preparing serial dilutions of high-sugar foods (Ingram 1957), the diluent should be adjusted to a low a_w value (Mossel & Sand 1968). Sterile glucose or sucrose (20%) solutions are suitable for this purpose, and should also be

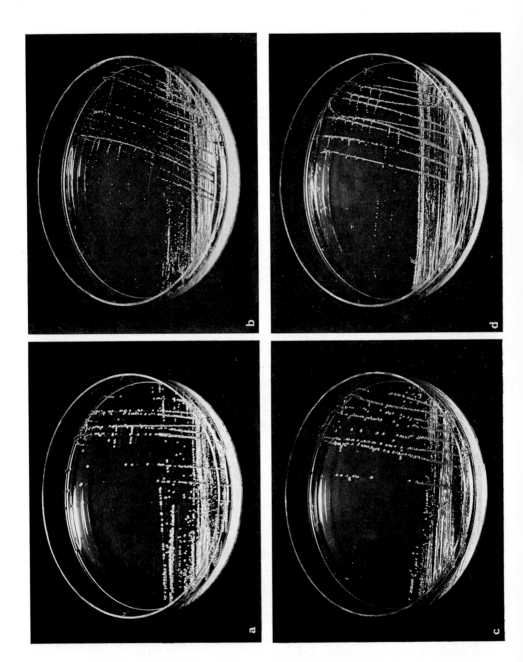

used for washing membranes. Secondly, where extended incubation periods of up to four weeks at 25–30°C are employed, drying out of the medium should be prevented by placing Petri dishes in closed boxes, polythene bags or desiccator jars.

Some osmophilic yeasts, particularly *S. bisporus* var. *mellis*, form clumps when growing in high-sugar foods. It was reported by Eddy (1955) that use of mannose or maltose (20%) diluent is effective in breaking up clumps of *S. cerevisiae*.

Total counts of osmophilic yeasts may be made by the standard microscopic technique using a haemocytometer at a magnification of ×400. It was observed by Scarr (1968) that osmophilic yeasts growing in concentrated sugar solutions vary in cell density with age and activity. She found that phase contrast examination of 60° Brix syrups enabled distinctions to be made between live, actively growing cells (phase bright), dead cells (phase dark) and moribund cells (grey). Use can be made of this technique to estimate the viable count and activity of osmophilic yeasts in consignments of raw sugar on arrival; this is a rapid way of indicating whether the sugar is likely to deteriorate on storage.

Maintenance and identification of osmophilic yeasts

Cultures are best maintained on high-sugar media, partly because some strains do not grow well on ordinary media like malt-extract agar (Sand 1973), and also because some strains may lose their tolerance to high sugar concentrations if kept on ordinary media (Onishi 1963; Scarr 1954). A suitable maintenance medium is osmophilic MYGP broth (Wickerham 1951): malt extract 3 g; yeast extract 3 g; peptone 5 g; glucose 500 g; distilled water to 1 l. pH is not adjusted. Sterilize by autoclaving at 115°C for 15 min. Cultures in this medium should be stored at 4°C and subcultured every six months. Some cultures may be freeze-dried successfully using the method of Rose (1970).

Identification of certain osmophilic yeasts, especially *S. rouxii* and *S. bisporus* var. *mellis*, is complicated by the fact that they usually exist in the haplophase and rarely produce ascospores (Davenport, pers. com. 1976). In recognition of this fact it is easier to omit some of the morphological tests described in Lodder (1970) and use instead the shorter identification schemes of Barnett & Pankhurst (1974) and Davenport (1974). Also, assimilation and fermentation patterns are best determined

Fig. 2. Colony morphology of osmotolerant yeasts on Scarr's osmophilic agar. Incubation conditions: 3 days, 30°C, aerobic. (a) *S. bailii* var. *osmophilus*; (b) *S. bisporus* var. *mellis*; (c) *S. rouxii*; (d) *T. versatilis*.

at sugar concentrations of 10% (w/v) rather than the usual 2% (Scarr & Rose 1965).

Stability and shelf-life tests

The rationale of such tests for confectionery products was discussed by Mossel & Sand (1968). Similar principles apply to testing of other high-sugar foods and will be mentioned here briefly in the light of experience gained in our laboratory. First, it is important to inoculate the test system with challenge organisms in an active physiological state. Ideally the inoculum should be prepared from cells actively growing in the same medium as the test system. Secondly, shelf-life tests should be of long duration due to the slow growth rate and extended lag phase of organisms growing at low a_w. Thirdly, cycling of incubation temperatures is desirable to simulate day and night temperatures, as this may cause localized a_w changes in the food. Fourthly, adequate numbers of sub-samples should be examined, as yeast growth may be unevenly distributed in the food. Clumping of yeast cells may also lead to falsely low counts; vigorous agitation of dilutions is therefore recommended. Finally, resuscitation of damaged cells is essential prior to their enumeration; the principles were outlined by Mossel (1975).

Occurrence of Food Spoilage by Osmophilic Yeasts

Many types of foods, raw materials or food ingredients rely upon reduced a_w to inhibit spoilage. Traditional methods of food preservation often achieved a 'safe' a_w by addition of sugars or by sun drying, whilst modern techniques such as vacuum evaporation achieve the same effect by concentration. Such foods, with a_w values in the range 0·60–0·85, are now grouped together as 'intermediate moisture foods' (Table 3). Although these products may be safe with respect to growth of food-borne pathogens (Leistner & Rodel 1976), they are susceptible to spoilage by xerotolerant yeasts and moulds (Tilbury 1976; Seiler 1976). The frequency and extent of these spoilage outbreaks and their economic effect is difficult to assess, due to commercial secrecy. Literature up to 1969 was comprehensively reviewed by Walker & Ayres (1970) whilst information since that date was surveyed by the author (Tilbury 1976). Table 4 summarizes some of the data from the latter paper, which should be consulted for more detail. Most of the unpublished results cited by the author arose from confidential work done on behalf of company customers. Similar data accumulated by Davenport (pers. comm. 1976) is presented in Table 5.

Osmophilic yeast spoilage of food usually becomes evident when counts

TABLE 3

*High-sugar content foods susceptible to spoilage by osmophilic yeasts (their
a_w levels and solute content, calculated as a proportion of the solute plus water)*

Product type	Examples	Solute content (% w/w)	Water activity a_w
(1) Syrups, sugars	Raw cane sugar	99·3–99·6	0·6–0·75
sweet spreads and	Refined sucrose syrup	66·7	0·85
preserves	Glucose or invert syrup	80	0·72
	Barley syrup; malt extract	75–80	0·7–0·8
	Maple syrup	64–74	0·7–0·8
	Honey, jam, marmalade	65–80	0·75–0·8
(2) Fruit juice	Orange juice	65	0·8–0·84
concentrates	Raspberry	65	0·79–0·8
(3) Confectionery	Marzipan	83–85	0·75–0·8
products	Glace cherries	70	0·75
	Toffees and caramels	92	0·6–0·65
(4) Bakery products	Fruit cakes	72–80	0·73–0·83
	Christmas pudding	75–80	0·7–0·77
(5) Dairy products	Sweetened condensed milk	70	0·83
(6) Dried fruits	Prunes and figs	80	0·68
	Dates	75–88	0·6–0·65

of 10^5–10^6 per ml or g are attained. Visible signs of fermentation are turbidity, gas evolution and swelling of containers, alcoholic and 'fruity' aromas, and discolouration (Fig. 3). Other effects of biodeterioration include loss of solids, increase in moisture content, and selective destruction of fructose in invert sugar (Tilbury 1967).

Effects of a_w on Growth of Osmophilic Yeasts

Growth rate

Many factors influence the effects of a_w on growth of osmophilic yeasts in high-sugar foods, but in general, their rate of growth is directly proportional to a_w. Hence the lower the a_w of a food the longer its microbiological shelf-life and the fewer the strains of yeasts able to grow in it.

Anand & Brown (1968) measured the growth rates of representative strains of osmophilic and non-osmophilic yeasts in a basal medium adjusted to various a_w levels with polyethylene glycol. None of the osmophilic yeasts showed a requirement for a reduced a_w, but they exhibited relatively broad a_w optima compared with the non-osmophilic

TABLE 4

Osmophilic yeast spoilage of some high-sugar foods (from Tilbury 1976)

High-sugar food	Predominant spoilage organisms	Reference
(1) Raw cane sugar	*S. rouxii*; *T. candida*	Tilbury (1967)
(2) Refined sugars and syrups	*T. apicola*; *T. globosa*	Scarr & Rose (1966)
	T. lactis-condensi	
	T. kestoni	
	C. guilliermondii	
	S. florentinus	
(3) Molasses	*S. heterogenicus*	
	T. holmii	Scarr (unpublished)
(4) Malt extract	*S. rouxii*	A. J. Reynolds (pers. comm., 1976)
(5) Fruit juice concentrates		Sand (1973)
(a) low a_w, no preservative	*S. rouxii*; *S. bisporus*	
(b) higher a_w, with preservative	*S. bailii*	
(6) Chocolate syrup	*T. etchellsii*	Tilbury (unpublished)
	T. versatilis	
	C. pelliculosa	
(7) Strawberry and apricot jams	*T. colliculosa*	Tilbury (unpublished)
	T. cantrellii	
(8) Jams	*S. bisporus*	Seiler (1976)
(9) Strawberry concentrate	*T. versatilis*	Tilbury (unpublished)
	C. pelliculosa; *C. utilis*	
(10) Crystallized and syruped ginger	*S. rouxii*	Lloyd (1975a)
(11) Comminuted orange base and cordial	*S. bailii* var. *bailii*	Lloyd (1975b)
(12) Fruit concentrates	*S. bailii* var. *bailii*	Pitt & Richardson (1973)
(13) Glace cherries	*S. rouxii*	Tilbury (unpublished)
(14) Marzipan and persipan	*S. rouxii*	Windisch &
	T. dattila	Neumann (1965)
(15) Soft-centred Easter eggs	*S. rouxii*	Tilbury (unpublished)

yeasts and had growth rates only half as fast as their respective optimum a_w values.

The author (Tilbury 1967) measured the growth rate of *S. rouxii* in sucrose broths whose a_w ranged from 0·997 to 0·935. Optimum growth occurred at an a_w of *ca.* 0·980 (25° Brix sucrose; mean generation time 1·30 h), whilst at a_w 0·935 (50° Brix sucrose) the mean generation time was 2·15 h. Growth rates and minimum a_w for growth were determined for 26 strains of osmophilic yeasts isolated from raw sugar, in sucrose/

TABLE 5

Occurrence of dominant spoilage yeasts in high-sugar products, Long Ashton Research Station, Bristol, 1969–1975. (Davenport pers. comm., 1976)

Organisms	Fruit concentrates (S.G. >1305)				Sugar syrups	
	Apple	Grape	Orange	Unknown	Brown	White
Osmophilic Yeasts						
Saccharomyces rouxii	+	+	+	+	+	+
S. bisporus var. *mellis*	−	+	+	−	−	−
S. bailii var. *osmophilus*	+	+	+	+	+	+
S. cerevisiae	+	+	+	−	−	−
S. bayanus	−	−	−	+	−	−
Saccharomyces spp.	+	+	+	+	+	+
Schizosaccharomyces pombe	−	−	+	−	−	−
T. versitalis	−	−	+	−	−	−
Osmotolerant Yeasts						
Candida valida	+	+	−	+	+	+
H. anomala var. *anomala*	+	+	−	+	+	+
Kloeckera apiculata	+	−	−	−	−	+
Candida spp.	+	+	−	+	−	+
Torulopsis spp.	+	+	+	+	−	+
S. cerevisiae	+	+	+	+	−	+
Saccharomyces spp.	+	+	+	+	+	+

Fig. 3. Microbial spoilage of packeted soft brown sugar. Discoloured patches are due to growth of xerotolerant yeasts.

TABLE 6

The effect of a_w on growth of osmophilic yeasts in sucrose/glycerol syrups after 12 weeks incubation at 27°C (Tilbury 1967)

Organism	Water activity a_w								Min. a_w for growth
	0·8	0·75	0·725	0·7	0·675	0·65	0·625	0·6	
S. rouxii (1)	NT	4	3	2	1	†	—	—	0·65
S. rouxii (2)	4	3	NT	2	NT	1	NT	—	0·65
S. bisporus var. mellis	4	3	NT	1	NT	—	NT	—	0·7
T. candida (1)	4	3	NT	1	NT	1	NT	—	0·65
T. candida (2)	NT	3	2	1	—	—	—	—	0·7
T. versatilis	NT	4	3	2	—	—	—	—	0·7
T. etchellsii	4	3	NT	†	NT	—	NT	—	0·7
H. anomala	2	†	NT	—	NT	—	NT	—	0·75

—, less than 1·5×original count; †, 1·5–2×original count; 1, 2–5×original count; 2, 5–10×original count; 3, 10–20×original count; 4, more than 20×original count; NT, not tested.

glycerol syrups incubated at 27°C for 12 weeks. Representative results are summarized in Table 6.

Horner & Anagnostopoulos (1973) measured the radial growth rate and lag phases of colonies of *S. rouxii* on sucrose agar over the a_w range 0·997 to 0·900. The lag phase for colony formation doubled to 40 h at a_w 0·900 compared with 20 h at a_w 0·960.

The implications of the low growth rates and extended lag phases of osmophilic yeasts at low a_w values are that high-sugar foods may have a 'safe' shelf-life of say, six months, but may spoil after 12 months. Once microbial growth begins, water is released, raising the a_w locally and accelerating the growth of the spoilage organisms (Mossel 1971; 1975).

Nature of the solute

Whilst a_w is a useful parameter for defining the effects of high solute concentrations on growth of osmophilic yeasts, it is now clear that the 'total solids' content of a food is better (Mossel 1975). At equal a_w values different solutes have different effects on growth. Anand & Brown (1968) showed that polyethylene glycol was more inhibitory than sucrose, whilst Horner & Anagnostopoulos (1973) found that glycerol was tolerated better than sucrose. Koppensteiner & Windisch (1971) determined the limiting values of osmotic pressure for growth of both osmophilic and non-osmophilic yeasts in solutions of a range of sugars and salts. The highest osmotic pressure tolerated (620 atm) occurred in fructose solutions; the salts were generally less well tolerated than the sugars. Salt tolerance increased in accordance with the Hofmeister series, i.e. NaCl was more inhibitory than KCl. Differences in the biological effects of humectants at equal a_w values were reported by Sinskey (1976). It may be concluded that the spoilage potential of a given food at a particular a_w value depends on the nature of the humectant and other food ingredients and should be determined by experiment.

Cell size and optical density

It was shown by Rose (1975) that the cell size of osmophilic yeasts decreased with reduction in a_w of the suspending medium, using sucrose and polyethylene glycol 200 as solutes. Differences were observed between solutes and also between these yeasts and non-osmophilic yeasts. It was postulated that osmotic pressure and cell permeability of solutes may exert specific effects independently of a_w. Similar general observations were made by Corry (1976), who showed by electron microscopy of freeze-etched preparations and from observations of increased optical density that plasmolysis occurred in S. rouxii cells in sucrose and sorbitol preparations, but less so in glucose, fructose and glycerol.

Other Factors Affecting Growth of Osmophilic Yeasts

Temperature

Osmophilic yeasts are similar to mesophilic yeasts in that they grow over the range 0–40°C with an optimum temperature of about 27°. In general, an increase in temperature increases the sugar concentration that a yeast can tolerate, and reduces the optimum a_w for growth (Ingram 1959). In the same way, both the optimum, minimum and maximum growth temperatures increase with increasing sugar concentration and reduction in a_w; this may explain why osmophilic spoilage, of dried fruits etc., is usually associated with hot climatic conditions (Ingram 1958). For example, S. rouxii and T. kestoni grow at 37°C on osmophilic agar (45% w/w sugar) but not on media containing 2% sugar (Scarr & Rose 1966). The phenomenon may explain the existence of apparently obligately osmophilic yeasts (Ingram 1957; Onishi 1963). Osmophilic yeasts probably behave like moulds in that they are most tolerant of low a_w near their optimum growth temperatures (Pitt 1975).

An important phenomenon occurs when a packaged food is stored at fluctuating temperatures. This results in moisture migration so that

certain parts of the food increase in a_w at the expense of others leading
to considerable local shifts in both growth rate and selection of microbial
types (Mossel 1975).

pH-Value

At high levels of a_w, the effects of pH, acidity and buffering capacity of
food on osmophilic yeasts are the same as for non-osmophilic yeasts, i.e.
growth usually occurs over the range pH 2·0–7·0 with an optimum pH
of 4·0–4·5. Hence the spoilage of medium and high acid foods, such as
fruit juice concentrates and pickles, is usually caused by yeasts rather
than by less acid-tolerant bacteria. It is thought that a reduction in pH
below the optimum reduces the maximum concentration of sugar tolerated; similarly, it is likely that a reduction in a_w of the food lessens the
tolerance of osmophilic yeasts to high acidity (Ingram 1959). Nevertheless, a few strains are able to tolerate both low a_w and high acidity, e.g.
fermentation of 65° Brix concentrated lemon juice at pH 2·0 by certain
Saccharomyces strains.

Redox potential (Eh) and gaseous environment

In a closed system, moisture vapour in the atmosphere rapidly equili-
brates with that derived from the stored food, until it reaches the
equilibrium relative humidity (ERH); this is numerically related to the
a_w of the food. Similarly, the oxygen content of the storage atmosphere
influences the redox potential (Eh) of the contained food, although the
former must usually be greatly changed before it affects the latter. The
Eh of a food depends on its redox poising capacity, which in turn is
governed by its content of reducing compounds, e.g. reducing enzymes,
reducing sugars, thiol-containing amino acids, ascorbic acid, etc. (Mossel
1971). Heat processing tends to raise the Eh of natural foods. The
majority of osmophilic yeasts are facultative anaerobes that grow most
rapidly under aerobic conditions, but they may also produce vigorous
fermentation under anaerobic conditions (Pitt 1975). An exception is
Debaryomyces hansenii which grows poorly in the absence of oxygen. The
ecological significance of this is that liquid foods are more susceptible to
spoilage by osmophilic yeasts than by xerophilic moulds, which, being
obligately aerobic, grow best near or on the surface of solid foods where
the Eh is higher. Osmophilic yeasts grow faster on the surface of solid
media than in liquid media; in fermenting liquids, growth is often confined to the superficial layers (Ingram 1959). Foods which possess both
low Eh and high acidity can only be spoilt by osmophilic yeasts. In
sealed containers fermentative spoilage by yeast may yield CO_2 which is
inhibitory to many aerobic organisms (Mossel 1971).

Chemical composition of the substrate

The chemical composition of a food affects its spoilage potential by osmophilic yeasts primarily with regard to the influence of 'total solids' on its a_w. A secondary factor is nutrient availability. In general, osmophilic yeasts are not fastidious in their nutritional requirements and they are able to grow on a wide range of foods. All osmophilic yeasts can utilize some simple sugars as a carbon source, e.g. glucose, fructose, maltose or sucrose, whilst some can also assimilate organic acids such as lactic and acetic; a few can hydrolyse starch. In contrast to many bacteria, yeasts are not proteolytic but can feebly attack a wide range of organic nitrogen compounds. Yeasts generally need only small amounts of nitrogenous compounds for growth, and are definitely fermentative rather than putrefactive. Consequently, foods with a high C/N ratio tend to be spoilt by yeasts, whereas foods with a high N/C ratio tend to be spoilt by bacteria. Similarly, yeasts need only small amounts of minerals for growth. As a group, yeasts possess good vitamin-synthesizing ability and hence are largely independent of an external supply (Ingram 1958); many osmophilic yeasts can grow in vitamin-free media. Natural antimicrobial substances may occur in foods and help to inhibit growth of spoilage yeasts, e.g. furfural and its derivatives produced in concentrated sugar syrups which have browned during storage (Ingram *et al.* 1955).

Physical state of the substrate

There is evidence that the structure and physical form of a food can influence the spoilage flora. Christian (1963) noted that osmophilic yeasts, in contrast to xerophilic moulds, may grow in liquid media at a_w values below 0·65, but not below a_w 0·75 on solid media. Possibly yeasts are less able than moulds to colonize solid substrates due to their lack of motility and non-mycelial mode of growth. An alternative explanation may be that the total water content is greater in liquids than in solids at the same a_w. It has been shown that total water content as well as a_w influences microbial growth response and that the minimal a_w for growth is higher in foods whose water content is adjusted by adsorption than by desorption (Acott & Labuza 1975). Additionally, oxygen tension varies with the substrate of a food. In nature, the combination of these factors results in the fact that spoilage of 'solid' foods like cereals, nuts and dried fruits tends to be caused by xerophilic moulds, whereas 'liquid' foods like fruit juice concentrates, syrups, jams and brines tend to be caused by osmophilic yeasts.

Competitiveness of osmophilic yeasts

High sugar content foods are such a selective environment that the relatively slow growth rate of osmophilic yeasts ceases to be a disadvantage. Yeasts possess great metabolic activity per cell, so that fermentation and visible spoilage may become evident at relatively low population levels. Their competitive advantages over moulds include their ability to grow anaerobically and produce and tolerate alcohol, CO_2 and organic acids (Ingram 1958). Their major competitive disadvantage compared with moulds may be the ability of the latter to penetrate and colonize solid substrates.

Survival and Prevention of Growth of Osmophilic Yeasts

Spoilage of high-sugar content foods by osmophilic yeasts can be prevented by heat sterilization, refrigeration or deep-freezing, but these processes are usually undesirable because of caramellization or high cost. More acceptable methods include elimination of contamination by good manufacturing practice ('GMP'); filtration; pasteurization; u.v. irradiation; use of humectants and acidulants; addition of food preservatives or combinations of several of these treatments.

Sources of contamination

Contamination of foods by osmophilic yeasts can arise by various means. Intrinsic infection of the food itself may occur, e.g. mummified fruits (Davenport 1975) but it is thought to be rare for this to be the sole important source of infection (Ingram 1958). Although yeasts are widely distributed in the air, osmophilic yeast infection is not normally airborne. Tilbury (1967) isolated small numbers of osmophilic yeasts in the air in raw cane sugar factories, but concluded that they did not significantly contaminate raw sugar. Scarr & Rose (1966) isolated a new species of osmophilic yeast, *Torulopsis kestoni* together with *Saccharomyces florentinus* from the air. Little is known about the occurrence of osmophilic yeasts in soil or water; Lochead & Farrell (1930) frequently isolated them from soil in an apiary, but Davenport (1975) found them to be relatively rare in vineyard soil. It is well known that insects such as bees, wasps and fruit-flies carry an indigenous population of osmophilic yeasts and that they have a significant role in transmission of yeasts to nectaries, fruits, honey, etc. (Ingram 1958; Lund 1958; Walker & Ayres 1970). In our laboratories *T. apicola*, *T. candida* and *S. rouxii* have been isolated from the abdominal contents of bees and wasps (Scarr &

Rose 1966). It is thought that such insects may help spread osmophilic yeasts in piles of raw sugar at refineries. Doubtless human carriers may also be important in certain instances, e.g. dried fruits (Ingram 1958). The major sources of osmophilic yeasts in foods are probably physical contact with already spoiled food or unclean equipment, e.g. residues of contaminated raw sugar on dirty processing plant (chutes, conveyors, scales, etc.) were shown to be the principal sources of contamination for raw cane sugar (Tilbury 1967).

Minimization of initial infection

Advantage may be taken of the slow growth rate and increased lag phase of osmophilic yeasts at low a_w values, to increase the shelf-life of products by minimizing the initial contamination. This can be achieved by 'GMP', including the use of good quality raw materials, plant sanitation, hygienic handling, prevention of delays in processing, exclusion of insects, safe packaging etc. Details of specific applications may be found for fruit juice concentrates (Sand 1973), raw sugar production (Tilbury 1967); jams (Horner & Anagnostopoulos 1973; Seiler 1976), confectionery (Mossel & Sand 1968) and bakery products (Seiler 1966, 1975).

In liquid products like liquid sugars, glucose syrups etc. osmophilic yeast counts can be reduced to less than one per ml by filtration through diatomaceous earth on a pre-coat filter. Concentration of products like liquid sugar, fruit juice, glucose syrups, malt extract, etc., by vacuum evaporation also acts as a pasteurization process, killing both ascosporogenous and non-ascosporogenous yeasts. Similar results are obtained where sugars are dissolved by heating during the manufacture of jams and confectionery products. Prevention of post-process contamination then becomes a key factor in pumping, storing, packaging and transporting the processed product.

Procedures adopted in the liquid sugar business were outlined by Scarr (1963). Pipes, tanks and vehicles are 'sterilized' by steaming until all surfaces in contact with sugar attain 80°C for at least 10 min. Microbiological air filters are fitted to storage tanks and both air and low-coloured liquid products may be sterilized by u.v. irradiation. It is essential to avoid condensation on the surface of stored liquid concentrates as it can dilute the top layers of the product and permit rapid growth of osmophilic yeasts. Condensation can be avoided by passing sterile warm air through the top of the tank.

Heat resistance of osmophilic yeasts

It was shown by Gibson (1973) using sucrose/glucose mixtures that a reduction in a_w of the heating menstruum markedly increases the heat resistance of osmophilic yeasts. More recent work by Corry (1976) demonstrated that at equal a_w values different solutes gave differing orders of protection, i.e. sucrose > sorbitol > glucose/fructose > glycerol. Some correlation was observed between the degree of plasmolysis in a solute and its degree of protectiveness. Obviously the nature and concentration of solute in a food quantitatively affects the degree of heat-processing needed to kill osmophilic yeast contaminants.

Put et al. (1976) determined the heat resistance of 120 strains of spoilage yeasts isolated from soft drinks and acid food products including some osmophilic yeasts. None survived heating for 20 min at 65°C at an inoculum level of ca. 10^5 cells in phosphate citrate buffer, pH 4·5, but ascosporogenous strains were generally more heat resistant than non-ascosporogenous strains. For comparison, Gibson (1973) obtained a maximum D-value of about 4·2 min for vegetative cells of T. globosa at a_w 0·72 and 65·5°C; on this basis 10^5 cells would have been killed after an exposure of about 21 min.

Research by Wilson et al. (1978) showed that the presence of sucrose or sodium chloride in the heating menstruum increased the heat resistance of vegetative cells of S. bailii and S. cerevisiae (osmotolerant strain). The heat resistance of released ascospores of S. cerevisiae was less affected by the presence of solutes than were the vegetative cells. A small proportion of ascospores of S. cerevisiae possessed considerably greater heat resistance at 55°C than the remainder or the vegetative cells of this organism. Neither vegetative cells of S. bailii and S. cerevisiae, nor ascospores of the latter strain were resistant to heating in acid conditions.

A practical illustration of the use of heat treatment to preserve high-sugar products was quoted by Lloyd (1975a) who 'pasteurized' syruped ginger by dipping it into 80° Brix sucrose syrup at 94±1°C for a minimum of 2 min. Effective 'flash' pasteurization processes used in the author's laboratory include 20 s at 85°C for a 50° Brix concentrated orange juice, and 30 s at 90°C for a 75° Brix chocolate flavoured syrup.

Antimicrobial preservatives

Many high-sugar foods are susceptible to spoilage because their a_w values are above the minimum for growth of xerotolerant yeasts and moulds. In products which are used periodically and which require a long shelf-life at ambient temperature (e.g. jams; sweet pickles; fruit drink concentrates), initial pasteurization or 'GMP' does not guarantee freedom from post-process contamination and subsequent spoilage. Here it is necessary to add antimicrobial preservatives, especially antimycotics.

Practical and legislative aspects of food preservatives were described by Jarvis & Burke (1976). Choice of antimycotics is restricted mainly to the following weak organic acids or their salts: benzoic acid; esters of parahydroxybenzoic acid; propionic acid and sorbic acid. Sulphurous acid in the form of SO_2, sulphite or bisulphite is widely permitted also. A problem in the application of these acid preservatives is that their antimicrobial activity is due to the undissociated acid and hence is strongly pH-dependent (Freese et al. 1973). Other weak organic acids such as acetic, citric and lactic acid are not classified as food preservatives but they are widely used in foods as acidulants. They exhibit some preservative activity which is primarily a function of their pH, although at the same pH value acetic acid is more inhibitory than lactic acid (Kimble 1977). There may be interactions between acidulants and other preservatives, e.g. lactic acid is synergistic with sorbate (Sahoo 1971).

In the UK, sulphite, benzoate and parahydroxybenzoate esters are the most widely permitted food preservatives. Within the EEC there is pressure to reduce permitted levels of SO_2 in foods, which has stimulated the search for effective alternatives, especially in fruit-based drinks (Ashworth & Jarvis 1975). It is recognized that SO_2 is not very effective against yeasts, especially in the presence of large amounts of reducing substances or where the pH exceeds 3·0. Combinations of SO_2 and benzoate are thought to be synergistic (Rehm & Wittmann 1962), and are used successfully to preserve low-acid foods and drinks against yeast spoilage. Sorbic acid is a better antimycotic than benzoic acid but at present its use in the UK is limited to certain bakery and confectionery products. Proposals to widen its application are likely to be approved (FAC 1977).

Benzoate alone or sorbate alone were found to be successful alternatives to SO_2 alone or SO_2 plus benzoate in the preservation of orange concentrate and orange drinks, against the osmophilic yeasts S. bailii, S. bisporus var. mellis, S. rouxii and Sch. pombe (Ashworth & Jarvis 1975). Further work confirmed that sorbate at 250–500 p.p.m. gave a similar or better shelf-life for fruit-based drinks than standard preservatives such as benzoic acid (550 p.p.m.) plus SO_2 (40 p.p.m.) (Patel 1977). In low-solids jam, however, sorbate at 125–250 p.p.m. inhibited xerophilic mould spoilage but did not prevent growth of S. bailii and Sch. pombe (Jarvis 1975).

It is well known that some spoilage yeasts develop tolerance to organic

124   R. H. TILBURY

acid preservatives (Warth 1977). The most resistant organism is the acid-tolerant and sugar-tolerant yeast, *S. bailii* (formerly *S. acidifaciens*). Ingram (1960*a,b*) described strains of *S. acidifaciens* resistant to 500 p.p.m. of benzoic acid in fruit squashes at pH 3·5. In spoilt fruit juice concentrates, Sand (1973) reported that 500 strains of *S. bailii* could tolerate 1000 p.p.m. of benzoic acid at pH 3·0, a_w 0·94. Lloyd (1975*b*) obtained similar results in Australian comminuted orange products, but sorbic acid inhibited *S. bailii* var. *bailii* at concentrations between 400 and 800 p.p.m. Use of sorbic acid, however, was precluded as it interfered with product colour; SO_2 at the maximum permitted level of 230 p.p.m. was found to be the only acceptable preservative, provided the pH was below 3·0. Pitt (1974) found that *S. bailii* was the most preservative-resistant spoilage yeast, tolerating 600 p.p.m. of both benzoic and sorbic acids at pH 2·5, in the presence of 10 % glucose. Other yeasts, including *C. krusei* and *T. holmii*, were inhibited by 300–400 p.p.m. of these preservatives.

Sugar and salt are traditional food preservatives but their use concentration is limited by their solubility and organoleptic acceptability. At use concentrations they may inhibit bacterial growth but only retard yeast and mould spoilage. These disadvantages have stimulated the search for new humectants in intermediate moisture foods (IMF's; Karel 1976). Ideally such compounds should have a bland flavour and possess antimicrobial properties in addition to their humectant role.

The polyhydric alcohols appear promising in these respects and already glycerol and propylene glycol are used commercially as humectants in human and pet foods respectively. A combination of propylene glycol (5–15 %) and potassium sorbate (0·1–0·2 %) is reported to be a most effective mycostat in IMF petfoods (Burrows & Barker 1976). Sinskey (1976) showed that other aliphatic diols or their esters were more effective than glycerol against bacteria but did not test osmophilic yeasts. However, the approval of regulatory authorities remains to be granted for these and many other promising new preservatives.

In conclusion, it is apparent that inhibition of growth of osmophilic yeasts in high-sugar foods is not achieved solely by control of one parameter such as a_w. Instead, reliance is placed on the 'hurdle' effect which utilizes a number of barriers to growth, such as pH, Eh, t-value, F-value and preservatives. The 'hurdle' concept is described in more detail by Leistner & Rodel (1976).

The author wishes to thank the Directors of Tate & Lyle Ltd. for their kind permission to publish this paper.

References

ACOTT, K. M. & LABUZA, T. P. 1975 Microbial growth response to water sorption preparation. *Journal of Food Technology* **10**, 603–611.

ANAGNOSTOPOULOS, G. D. 1973 Water activity in biological systems: a dew-point method for its determination. *Journal of General Microbiology* **77**, 233–235.

ANAND, J. C. 1969 The Physiology and Biochemistry of Sugar-Tolerant Yeasts. *Ph.D. Thesis*, University of New South Wales.

ANAND, J. C. & BROWN, A. D. 1968 Growth rate patterns of the so-called osmophilic and non-osmophilic yeasts in solutions of polyethylene glycol. *Journal of General Microbiology* **52**, 205–212.

ASHWORTH, J. & JARVIS, B. 1975 The replacement of sulphur dioxide as the microbial preservative in orange concentrate and orange drinks. *British Food Manufacturing Industries Research Association Research Report* No. 227.

AYERST, G. 1965 Water activity—Its measurement and significance in biology. *International Biodeterioration Bulletin* **1**, 13–26.

BARNETT, J. A. & PANKHURST, R. J. 1974 *A New Key to the Yeasts* Amsterdam, London: North-Holland Publishing Co.

BONE, D. 1973 Water activity in intermediate moisture foods. *Food Technology* **27** (4), 71–76.

BROWN, A. D. 1976 Microbial water stress. *Bacteriological Reviews* **40**, 803–846.

BURROWS, I. E. & BARKER, D. 1976 Intermediate moisture petfoods. In *Intermediate Moisture Foods* eds Davies, R., Birch, G. G. & Parker, K. J. Ch. 4, pp. 43–53, London: Applied Science.

CHRISTIAN, J. H. B. 1963 Water activity and the growth of micro-organisms. In *Recent Advances in Food Science* eds Leitch, J. M. & Rhodes, D. N. Vol. 3, pp. 248–255, London: Butterworths.

CORRY, J. E. L. 1974. The Effect of Sugars and Polyols on the Heat Resistance of Salmonellae and Osmophilic Yeasts. *Ph.D. Thesis*, University of Surrey.

CORRY, J. E. L. 1976 The effect of sugars and polyols on the heat resistance and morphology of osmophilic yeasts. *Journal of Applied Bacteriology* **40, 269–276**.

DAVENPORT, R. R. 1974 A simple method, using stripdex equipment, for the assessment of yeast taxonomic data and identification keys. *Journal of Applied Bacteriology* **37**, 269–271.

DAVENPORT, R. R. 1975. The Distribution of Yeasts and Yeastlike Organisms in an English Vineyard. *Ph.D. Thesis*, University of Bristol.

EDDY, A. A. 1955 Floculation characteristics of yeasts II. Sugars as dispersing agents. *Journal of the Institute of Brewing* **61**, 313–317.

FAC/Rep/24, 1977 Food additives and contaminants committee report on the review of the use of sorbic acid in food. London: HMSO.

FREESE, E., SHEU, C. W. & GALLIERS, E. 1973 Function of lipophilic acids as antimicrobial food additives. *Nature, London* **241**, 321–325.

GIBSON, B. 1973 The effect of high sugar concentrations on the heat resistance of vegetative micro-organisms. *Journal of Applied Bacteriology* **36**, 265–276.

HARDMAN, T. M. 1976 Measurement of water activity: Critical appraisal of methods. In *Intermediate Moisture Foods* eds Davies, R., Birch, G. G. & Parker, K.J. Ch. 7, pp. 75–88, London: Applied Science.

HORNER, K. J. & ANAGNOSTOPOULOS, G. D. 1973 Combined effects of water activity, pH and temperature on the growth and spoilage potential of fungi. *Journal of Applied Bacteriology* **36**, 427–436.

INGRAM, M. 1957 Micro-organisms resisting high concentration of sugars or salts.

In *Microbial Ecology, Symposium of the Society for General Microbiology* eds Pollock, M. R. & Richmond, M. H. Vol. 7, pp. 90–133. Cambridge: University Press.

INGRAM, M. 1958 Yeasts in food spoilage. In *The Chemistry and Biology of Yeasts* ed. Cook, A. H. pp. 603–633, London: Academic Press.

INGRAM, M. 1959 Physiological properties of osmophilic yeasts. *Review Fermentation Industrie Alimentaria* **14**, 23–33.

INGRAM, M. 1960a Studies on benzoate-resistant yeasts. *Acta Microbiologica* **7**, 95–105.

INGRAM, M. 1960b An influence of carbon source on the resistance of a yeast to benzoic acid. *Annals Institut Pasteur* **11**, 167–178.

INGRAM, M., MOSSEL, D. A. A. & DE LANGE, P. 1955 Factors produced in sugar-acid browning reactions, which inhibit fermentation. *Chemistry and Industry* p. 63.

JARVIS, B. 1975 Sorbic acid as a preservative for 'low solids' jam. *British Food Manufacturing Industries Research Association Technical Circular* No. 594.

JARVIS, B. & BURKE, C. S. 1976 Practical and legislative aspects of the chemical preservation of foods. In *Inhibition and Inactivation of Vegetative Microbes* eds Skinner, F. A. & Hugo, W. B. pp. 345–367, London: Academic Press.

KAREL, M. 1976 Technology and application of new intermediate moisture foods. In *Intermediate Moisture Foods* eds Davies, R., Birch, G. G. & Parker, K. J. Ch. 2, pp. 4–31, London: Applied Science.

KIMBLE, C. E. 1977 Chemical food preservatives. In *Disinfection, Sterilisation and Preservation* 2nd edn, ed. Block, S. S. Ch. 41, pp. 834–858 Philadelphia: Lea & Febiger.

KOH, T. Y. 1972 Studies on the Osmophilic Yeast *Saccharomyces rouxii*. Ph.D. Thesis, University of Cambridge.

KOPPENSTEINER, G. & WINDISCH, S. 1971 Osmotic pressure as a limiting factor for growth and fermentation of yeasts. *Archiv für Mikrobiologie* **80**, 300–314.

KUSHNER, D. J. 1971 Influence of solutes and ions on micro-organisms. In *Inhibition and Destruction of the Microbial Cell* ed. Hugo, W. B. Ch. 5, pp. 259–283, London: Academic Press.

LANDROCK, A. H. & PROCTOR, B. E. 1951 Measuring humidity equilibrium. *Modern Packaging* **24**, 123–130 and 186.

LEISTNER, L. & RÖDEL, W. 1976 The stability of intermediate moisture foods with respect to micro-organisms. In *Intermediate Moisture Foods* eds. Davies, R., Birch, G. G. & Parker, K. J. Ch. 10, pp. 120–137, London: Applied Science.

LLOYD, A. C. 1975a Osmophilic yeasts in preserved ginger products. *Journal of Food Technology* **10**, 575–581.

LLOYD, A. C. 1975b Preservation of comminuted orange products. *Journal of Food Technology* **10**, 565–574.

LOCHEAD, A. G. & FARRELL, L. 1930 Soil as a source of infection of honey by sugar-tolerant yeasts. *Canadian Journal of Research* **3**, 51–64.

LODDER, J. (ed.) 1970 *The Yeasts: A Taxonomic Study* Amsterdam, London: North-Holland Publishing Co.

LUND, A. 1958 Ecology of yeasts. In *The Chemistry and Biology of Yeasts* ed. Cook, A. H. pp. 63–91, London: Academic Press.

MOSSEL, D. A. A. 1951 Investigation of a case of fermentation in fruit products rich in sugar. *Antonie van Leeuwenhoek* **17**, 146–152.

MOSSEL, D. A. A. 1971 Physiological and metabolic attributes of microbial groups associated with foods. *Journal of Applied Bacteriology* **34**, 95–118.

Mossel, D. A. A. 1975 Water and micro-organisms in foods—a synthesis. In *Water Relations of Foods* ed. Duckworth, R. pp. 347–361, London: Academic Press.

Mossel, D. A. A. & Sand, F. E. M. J. 1968 Occurrence and prevention of microbial deterioration of confectionery products. *Conserva* 17, 23–32.

Norrish, R. S. 1966 An equation for the activity coefficients and equilibrium relative humidities of water in confectionery syrups. *Journal of Food Technology* 1, 25–39.

Onishi, H. 1963 Osmophilic yeasts. *Advances in Food Research* 12, 53–94.

Patel, M. 1977 Use of sorbic acid as a preservative in fruit-based drinks. *British Food Manufacturing Industries Research Association Technical Circular* No. 645.

Pitt, J. I. 1974 Resistance of some food spoilage yeasts to preservatives. *Food Technology (Australia)* 26, 238–241.

Pitt, J. I. 1975 Xerophilic fungi and the spoilage of foods of plant origin. In *Water Relations of Foods* ed. R. Duckworth, pp. 273–307, London: Academic Press.

Pitt, J. I. & Richardson, K. C. 1973 Spoilage by preservative-resistant yeasts. CSIRO *Food Research Quarterly* 33, 80–85.

Put, H. M. C., de Jong, J., Sand, F. E. M. J. & van Grinsven, A. M. 1976 Heat resistance studies on yeast spp. causing spoilage in soft drinks. *Journal of Applied Bacteriology* 40, 135–152.

Rehm, H. J. & Wittmann, H. 1962 Beitrag zur Kenntnis der Antimicrobielle Wirkung der Schwefligen Saure I Übersicht über Einflussnehmende Faktorenauf die Antimikrobielle Wirkung der Schwefligen Saure. *Zeitschrift für Lebensmittel Forsch* 118, 413–429.

Robinson, R. A. & Stokes, R. H. 1965 *Electrolyte Solutions* 2nd edn, London: Butterworths.

Rose, D. 1970 Some factors influencing the survival of freeze-dried yeast cultures. *Journal of Applied Bacteriology* 33, 228–232.

Rose, D. 1975 Physical responses of yeast cells to osmotic shock. *Journal of Applied Bacteriology* 38, 169–175.

Ross, K. D. 1975 Estimation of water activity in intermediate moisture foods. *Food Technology* 29, (3), 26–34.

Sahoo, B. N. 1971 *Ph.D. Thesis*: University of Missouri. Cited by Sinskey, A. J. 1976 In *Intermediate Moisture Foods* eds Davies, R., Birch, G. G. & Parker, K. J. Ch. 18, p. 262, London: Applied Science.

Sand, F. E. M. J. 1973 Recent investigations on the microbiology of fruit juice concentrates. In *Technology of Fruit Juice Concentrates—Chemical Composition of Fruit Juices*, pp. 185–216 International Federation of Fruit Juice Producers, Scientific-Technical Commission 13, Vienna.

Scarr, M. P. 1954 Studies on the Taxonomy and Physiology of Osmophilic Yeasts Isolated from the Sugar Cane. *Ph.D. Thesis*, University of London.

Scarr, M. P. 1959 Selective media used in the microbiological examination of sugar products. *Journal of the Science of Food and Agriculture* 10, 678–681.

Scarr, M. P. 1963 Microbiological standards for sugar. *Journal of the Science of Food and Agriculture* 14, 220–223.

Scarr, M. P. 1968 Studies arising from observations of osmophilic yeasts by phase contrast microscopy. *Journal of Applied Bacteriology* 31, 525–529.

Scarr, M. P. & Rose, D. 1965 Assimilation and fermentation patterns of osmophilic yeasts in sugar broths at two concentrations. *Nature, London* 207, 887.

SCARR, M. P. & ROSE, D. 1966 Study of osmophilic yeasts producing invertase. *Journal of General Microbiology* **45**, 9–16.

SCOTT, W. J. 1957 Water relations of food spoilage micro-organisms. *Advances in Food Research* **7**, 83–127.

SEILER, D. A. L. 1966 Fermentation problems in high sugar coatings and fillings. *British Baking Industry Research Association Bulletin* No. 2 April, 49–53.

SEILER, D. A. L. 1975 Fermentation of fudge and fondant *British Baking Industry Research Association Bulletin* No. 2, April, 58–60.

SEILER, D. A. L. 1976 The stability of intermediate moisture foods with respect to mould growth. In *Intermediate Moisture Foods* eds Davies, R., Birch, G. G. & Parker, K. J. Ch. 12, pp. 166–181, London: Applied Science.

SINSKEY, A. J. 1976 New developments in intermediate moisture foods: humectants. In *Intermediate Moisture Foods* eds Davies, R., Birch, G. G. & Parker, K. J. Ch. 18, pp. 260–280, London: Applied Science.

TILBURY, R. H. 1967 Studies on the Microbiological Deterioration of Raw Cane Sugar. *M.Sc. Thesis*, University of Bristol.

TILBURY, R. H. 1976 The microbial stability of intermediate moisture foods with respect to yeasts. In *Intermediate Moisture Foods* eds Davies, R., Birch, G. G. & Parker, K. J. Ch. 11, pp. 138–165, London: Applied Science.

TROLLER, J. A. 1977 Statistical analysis of a_w measurements obtained with the Sina Scope. *Journal of Food Science* **42**, 86–90.

VAN DER WALT, J. P. 1970 Criteria and methods used in classification. In *The Yeasts: A Taxonomic Study* ed. Lodder, J. Ch. 2, pp. 34–113, Amsterdam, London: North-Holland Publishing Co.

WALKER, H. W. & AYRES, J. C. 1970 Yeasts as spoilage organisms. In *The Yeasts* eds Rose, A. H. & Harrison, J. S. Vol. 3, Yeast Technology, Ch. 9, pp. 463–527, London: Academic Press.

WARTH, A. D. 1977 Mechanism of resistance of *Saccharomyces bailii* to benzoic, sorbic and other weak acids used as food preservatives. *Journal of Applied Bacteriology* **43**, 215–230.

WICKERHAM, L. J. 1951 *The Taxonomy of Yeasts*. Tech. Bull. No. 1029, US Department of Agriculture, Washington, DC.

WILSON, J. M., WOOD, J. M. & JARVIS, B. 1978 The heat resistance of vegetative and ascospore forms of yeasts in relation to their environmental conditions. *British Food Manufacturing Industries Research Association Research Report* No. 275.

WINDISCH, S. & NEUMANN, I. 1965 Zur Mikrobiologischen Untersuchung bon Marzipan. *Susswaren*, **7**, 355–358, 484–490, 540–546.

Spoilage of Materials of Reduced Water Activity by Xerophilic Fungi

H. DALLYN

Metal Box Ltd., Denchworth Road, Wantage, UK

A. FOX

Department of Applied Biology and Food Science, Polytechnic of the South Bank, London, UK

Fungi are ubiquitous, and a number of species are able to tolerate very dry conditions. The types possessing this characteristic are referred to as xerophiles, the term being used to imply the ability of the fungus to tolerate conditions of low water activity (a_w), rather than in the obligate sense. Some of the more xerophilic species, however, are unable to grow at high a_w values, and the growth of others is much restricted (Scott 1957).

Pitt (1975) defines xerophilic fungi as those able to germinate and grow at an a_w value of 0·85. Any attempt to fix precise values of a_w for the organisms has limitations, but that suggested by Pitt would seem a suitable compromise. Growth optima for xerophiles, although reduced, still lie near or above, a_w 0·90 (usually 0·90–0·97). Growth of xerophiles can be slow even at their optima, so that it is advisable to use media of low a_w to eliminate overgrowth by more vigorously growing fungi.

Unprocessed foods and materials may carry a wide variety of micro-fungal spores, and although storage conditions will universally affect out-growth, not all of them will have potential as spoilage organisms. Mitchell & Stauber (1975) outline the spoilage problem areas for tobacco. During the various stages of production there is a succession from species growing at the high a_w values of newly harvested leaves, through to those tolerating the low a_w values of the final, cured product. Foods and materials preserved by drying are likely to be subject to a similar succession of potential spoilage fungi, and each of these may be present on the final product. The microbial stability of the product will depend, among other things, on its a_w value and the presence of inhibitory systems.

When spoilage does occur in a dried food or product, then the agent is commonly one or more species of xerophilic fungi. Typical examples are shown in Figs 1 and 2. Fig. 1 shows spoilage of dried paw-paw (a_w 0·78); the obvious spoilage organism was shown to be *Wallemia sebi* but a secondary growth of *Monascus bisporus* was also found on culturing. Figure 2 shows spoilage of prunes by *M. bisporus*. This paper gives methods for the isolation and study of the growth of xerophilic fungi.

Fig. 1. Growth of *W. sebi* on dried West Indian paw-paw.

Fig. 2. Dried prunes inoculated with *M. bisporus*.

Isolation Methods

Media

The general principle involves the use of media with the a_w reduced to the point at which the growth rate of species of non-xerophiles is much retarded or where growth is prevented altogether. Media with a_w values of about 0·90 are very useful in that they retard the growth rate of most species including many of the more prolific xerophiles. Thus detection and isolation of the less vigorously growing xerophilic species is possible.

Pitt (1975) discusses the merits of some of the media devised for growing xerophilic species of fungi. As a general medium however, malt extract agar with the a_w adjusted to 0·90–0·92 with glycerol will support the growth of most species encountered. This can be made up with granules (Oxoid) or, if preferred, from the basic ingredients g l^{-1}: malt extract, 30; mycological peptone (or suitable substitute), 5; Oxoid agar no. 1, 15; 4·5 molal aqueous glycerol solution (instead of water), 1 l; pH 5·4 approximately (Galloway & Burgess 1950).

The a_w of the glycerol solution can be adjusted using the values given in Table 1 (the concentration shown in the formulation should give a medium with a final a_w value of 0·91–0·92).

TABLE 1

The relationship of molality to a_w of aqueous glycerol solutions (from Dallyn 1978)

Molality†	a_w	Molality	a_w	Molality	a_w
0·1	0·998	2·5	0·955	10	0·825
0·2	0·996	3·0	0·946	11	0·809
0·3	0·995	3·5	0·937	12	0·793
0·4	0·993	4·0	0·928	13	0·777
0·5	0·991	4·5	0·919	14	0·762
0·6	0·989	5·0	0·910	15	0·747
0·7	0·987	5·5	0·901	16	0·733
0·8	0·986	6·0	0·892	17	0·719
0·9	0·983	6·5	0·884	18	0·705
1·0	0·982	7·0	0·875	19	0·691
1·2	0·978	7·5	0·867	20	0·678
1·4	0·975	8·0	0·858	22	0·650*
0·6	0·971	8·5	0·850	23·75	0·625*
1·8	0·967	9·0	0·841	25·50	0·600*
2·0	0·964	9·5	0·833		

* Obtained by extrapolation.
† Molality is defined as moles of solute kg—1 of solvent.

Sucrose or glucose are sometimes used instead of glycerol as these have the advantage of being relatively cheap. The disadvantage of using sucrose is that excessive heating can result in some inversion of the sucrose with subsequent change in a_w. Data for sucrose is given by Scott (1957) whilst some data for glucose is given by Reid (1976).

The use of glucose has been recommended by Pitt (1975) in his MY40G and MY60G formulations. MY60G is said to have an a_w approximating to 0·85 so that it is useful as a means of establishing the presence of xerophiles. The two formulations given by Pitt are as follows:

(1) MY40G (malt extract, yeast extract 40% glucose agar): malt extract, 12 g; yeast extract, 3 g; glucose, 400 g; agar, 12 g; water, 600 g. Sterilize by steaming for 30 min. The final pH is about 5·5 and the a_w is near 0·92.

(2) MY60G (malt extract, yeast extract 60% glucose agar): malt extract, 8 g; yeast extract, 2 g; glucose, 600 g; agar, 8 g; water, 400 g. Sterilize by steaming for 30 min.

Other useful media include Czapek-Dox agar with increased sucrose concentrations as recommended by Raper & Fennell (1965). Malt extract, yeast extract, glucose, peptone agar (MYGP agar) described by Barr & Downey (1975) for use in detecting mycotoxins can be adapted for growing *Monascus*, *Eremascus* and *Bettsia* if the a_w value of the medium is suitably adjusted.

Sterilization of media

To reduce non-enzymic browning reactions, sterilization treatment should be reduced to as low a level as practiable. Media with a_w values of 0·92 or below can be sterilized by steaming at 100°C for 30 min. Media with a_w values above 0·92 require a pressure process: 115°C for 10 min is satisfactory where the pH is below 5·5, but 121°C for 15 min is necessary for higher pH values.

Isolation

Materials with obvious fungal spoilage should be examined using low-power microscopy followed by high-power microscopic scrutiny of suitable, mounted material. Identification of the species present and other important information may be obtained at this stage.

Small pieces of material excised with a sterile scalpel should be placed directly onto prepared plates of a suitable low a_w agar medium, or imprints from the material may be taken directly onto the agar surface.

In order to prevent dehydration of media during the necessary long incubation, Petri dish cultures may be sealed with Parafilm (Gallenkamp Ltd.), or placed in a low density polyethylene bag or an airtight container (e.g. a tin with a tight-fitting lid). Alternatively cultures may be held in a desiccator containing a prepared solution with an a_w value the same as that of the medium.

Identification

The keys given by Pitt (1975) form a useful basis for identification, but do not include *Bettsia alvei*, an extreme xerophile which occasionally causes spoilage in some honey products (Dallyn 1978). *B. alvei* forms pure white colonies on most rich, low a_w media. It produces aleuriospores and chlamydospores, and given favourable conditions may produce microcysts with round ascospores. The fungus is capable of growth from spores at least down to a_w values of 0·71, but on normal culture media with an a_w of 0·99 growth is very restricted or absent. A full description of *B. alvei* is given by Skou (1972).

Many of the xerophilic species have been described by Pitt (1975); further information can be obtained from Raper & Fennell (1965) and Smith (1969). The latter volume is under revision. The scanning electron microscope can be a valuable tool for establishing the identity of members of the *A. glaucus* group (Locci 1972). Examples are shown in Figs 3 and 4.

Maintenance of Cultures

Some fungus cultures can be hazardous, and staff need to be adequately trained in the skills of handling micro-organisms. Rubidge & Austin (1977) describe some of the hazards, and recommend various precautions when testing the fungus resistance of materials, including the use of a safety cabinet of the exhaust type to protect the operator.

Cultures are grown on a medium which induces maximum spore formation such as one of those listed above. The choice of medium has to be determined by experiment as different species have their own requirements. In the authors' experience malt extract agar with the a_w adjusted to 0·90–0·92 with either glycerol or sucrose has been found to be satisfactory for the majority of species. Normal methods for storing cultures are detailed by Smith (1969) and Codner (1972). A particularly convenient method is to store spores of cultures in sterile bentonite. Bentonite is sterilized by heating in a hot air oven at 160°C for 2 h in 0·25 oz Bijou bottles plugged with cotton wool. The cotton wool plug can be replaced by a metal cap (which is fitted with a rubber liner) and has been pre-sterilized by autoclaving. Spores and bentonite can be

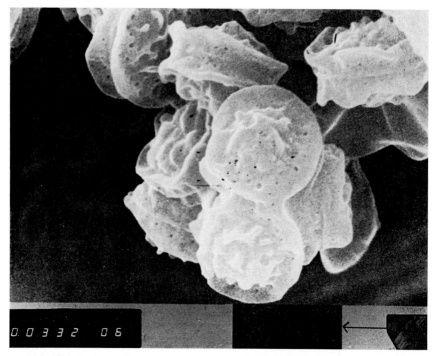

Fig. 3. Scanning electron micrograph of *A. cristatus* ascospores.

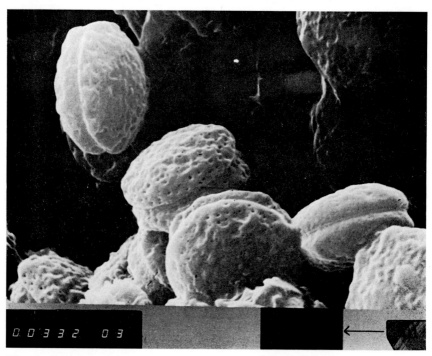

Fig. 4. Scanning electron micrograph of unidentified large-spored *A. glaucus* group ascospores

mixed by shaking and the resulting dry suspension used for inoculation studies. Bentonite suspensions remain stable for some time; in the author's laboratory (H.D.) spores of *Aspergillus niger* harvested and stored in this manner have been shown to retain their viability after 20 years at room temperature. Storage in bentonite markedly reduces aerosol formation and constitutes a useful source of spore inocula. Mixed fungal spore suspensions of appropriate strains may be used for inoculating foods or materials in mould susceptibility tests. Application of spore suspensions can be made with a camel hair brush.

Determination of Growth Rate

Dry weight determinations have been found to give a satisfactory measure of growth rate at low substrate a_w values. The following method has been used successfully.

Preparation of inoculum

Dry mould spore suspensions prepared in sterile bentonite, as described above, serve as the inoculum. By careful mixing it is possible to obtain a relatively homogeneous suspension. The age of the spores can affect germination, and for consistently reproducible germination, newly harvested but mature spores have been found to be best.

Preparation of substrate

A range of agar media of differing a_w values can be prepared with glycerol. Other humectants can be used, but glycerol is one which has been found to give least inhibitory effects although the choice may be dictated by a particular organism or situation.

Using the relationship given by Scott (1957):

$$a_w = \exp\left(-\frac{vm\phi}{55\cdot51}\right)$$

where ϕ is the molal osmotic coefficient, v is the number of ions generated by each molecule (for non-electrolytes $v = 1$), and m is the molality of the solution. Using values for ϕ obtained by Scatchard *et al.* (1938) the a_w values given in Table 1 can be calculated. Hence aqueous solutions of glycerol of the required a_w values can be prepared, and using a fixed volume of these solutions media can be prepared to cover the desired a_w range.

The a_w value of each medium can be checked using a suitable method. Troller & Christian (1978) give a summary of methods for controlling and determining a_w as well as an extensive list of references. The Sina apparatus (Fig. 5) described by Troller & Christian has been found to

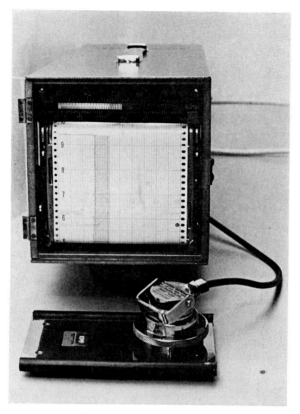

Fig. 5. Sina apparatus used for measuring a_w. The sensor (normally housed in constant temperature cabinet) is shown in the foreground.

give consistent readings providing sufficient care is exercised. The instrument is described as an electrolytic hygrometer. For maximum accuracy the sensor head should be housed in a temperature controlled cabinet. The instrument needs careful calibration against known standards (e.g. those given by Robinson & Stokes (1955) and Young (1967)) so that a calibration curve can be constructed. Corrections can then be applied to unknown samples. It is often convenient to store the standard saturated salt solutions in the same temperature controlled cabinet as the sensor to reduce the time needed for temperature equilibration.

An additional method for rapid a_w determinations of a particular medium containing glycerol can be devised if the refractive index of the samples is measured with an Abbé refractometer as well as determining the a_w values with the Sina apparatus (Nova-Sina Ltd., Zurich, Switzerland). A curve relating a_w values with refractive indices can be constructed and subsequently used for rapid a_w checks on that particular medium.

In addition, Troller & Christian (1978) give data on the relationship between refractive indices and concentrations for a range of glycerol solutions. Further information on this is given by the United Kingdom Glycerine Producers' Association.

Preparation and inoculation of Petri dish cultures

Petri dishes should be poured with the prepared medium to a standard depth. For low a_w media where growth is limited, dishes divided into quadrants are convenient. Uncoated sterile cellophane discs (code 325P supplied by D. J. Parry and Co. Ltd., 7 Avon Trading Estate, London W14 8UE manufactured by British Cellophane, Bridgwater, Somerset) 20 μm thick are placed onto the agar surface using sterile forceps. The size of the discs needed will depend on the growth rate. For low a_w values 30 mm diam. discs are satisfactory, but larger ones, slightly less than the diameter of the agar surface in the Petri dish, are needed for higher a_w values. In this manner a series of Petri dishes containing media, overlaid with cellophane discs, can be prepared. Sufficient plates are prepared so that replicate dry weight determinations can be made at selected intervals during the growth of the fungus.

The bentonite suspension can be applied to the centre of each cellophane disc using a sterile applicator stick. This will result in a small round area (*ca.* 2·0 mm in diameter) of inoculum comprising bentonite spore suspension. It is preferable to lower the prepared dish onto the stick. The inoculated dishes can be incubated in sealed containers (e.g. slip lid polish tins 11·7 cm diameter, screw lid tins of the type for packaging boiled sweets or, alternatively, heat-sealed polyethylene bags).

The inoculated Petri dishes need to be incubated at constant temperature and examined at intervals for spore germination. A record can be kept of the time taken for the spores to germinate. This will be different for each a_w and species. As the mycelium develops, replicate mats can be harvested, dried and weighed at regular intervals. Figure 6 shows the growth rate pattern for *Eremascus albus*. The method could provide valuable information on the growth rate of fungi under a particular regime. Additionally it could be adapted to study the effect of a_w on the formation of secondary metabolites such as mycotoxins.

The authors are most grateful to Mr R. J. Whittaker for preparing the scanning electron micrograph and the photographs and to Miss J. King for help in assembling the paper and to Dr A. H. S. Onions for assistance in identification.

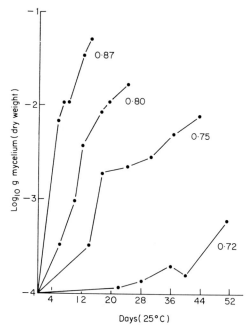

Fig. 6. Growth of *E. albus* at various a_w values on malt extract agar, a_w adjusted with glycerol (from Dallyn 1978).

References

BARR, J. G. & DOWNEY, G. A. 1975 A multiple inoculation technique for the screening of fungal isolates for the evaluation of growth and mycotoxin production on agar substrates. *Journal of the Science of Food and Agriculture* **26,** 1561–1566.

CODNER, R. C. 1972 Preservation of fungal cultures and the control of mycophagous mites. In *Safety in Microbiology* eds Shapton, D. A. & Board, R. G. *Society for Applied Bacteriology, Technical Series* No. 6, pp. 213–227, London and New York: Academic Press.

DALLYN, H. 1978 The Effect of Substrate Water Activity on the Growth of Certain Xerophilic Fungi. *Ph.D. Thesis*, CNAA Polytechnic of the South Bank, London.

GALLOWAY, L. D. & BURGESS, R. 1950 *Applied Mycology and Bacteriology* 3rd edn, London: Leonard Hill Ltd.

LOCCI, R. 1972 Scanning electron microscopy of ascosporic Aspergilli. *Supplimento al volume 8 (series 4) della Rivista di Patologia Vegetale.*

MITCHELL, T. G. & STAUBER, P. C. 1975 Biodeterioration of tobacco. In *Microbial Aspects of the Deterioration of Materials* eds Lovelock, D. W. & Gilbert, R. J. *Society for Applied Bacteriology, Technical Series* No. 9, pp. 203–211, London and New York: Academic Press.

PITT, J. I. 1975 Xerophilic fungi and the spoilage of foods of plant origin. In *Water Relations of Foods* ed. Duckworth, R. B. pp. 273–307, London and New York: Academic Press.

RAPER, K. B. & FENNELL, D. I. 1965 *The Genus* Aspergillus. Baltimore: Williams & Wilkins.

REID, D. S. 1976 Water activity concepts in intermediate moisture foods. In *Intermediate Moisture Foods* eds Davies, R., Birch, G. G. and Parker, K. J. p. 57, London: Applied Science Publishers.

ROBINSON, R. A. & STOKES, R. H. 1955 *Electrolyte Solutions* London: Butterworths.

RUBIDGE, T. & AUSTIN, I. E. 1977 Fungus testing of materials. *Journal of Society of Environmental Engineering* September, 16–18.

SCATCHARD, G., HAMER, W. J. & WOOD, S. E. 1938 Isotonic solutions. I. The chemical potential of water in aqueous solutions of sodium chloride, potassium chloride, sulphuric acid, sucrose, urea and glycerol at 25°C. *Journal of the American Chemical Society* **60,** 3061–3070.

SCOTT, W. J. 1957 Water relations of food spoilage micro-organisms. In *Advances in Food Research* eds Mrak, E. M. & Stewart, G. F. Vol. 7, pp. 83–127, New York and London: Academic Press.

SKOU, J. P. 1972 Ascosphaerales. *Friesia* **10,** 1–24.

SMITH, G. 1969 *An Introduction to Industrial Mycology* 6th edn, London: Arnold.

TROLLER, J. A. & CHRISTIAN, J. H. B. 1978 *Water Activity and Food* New York & London: Academic Press.

UNITED KINGDOM GLYCERINE PRODUCERS' ASSOCIATION (undated) *Glycerine, its Physical Properties*. Pool Lane, Bebington, Wirral, Merseyside L62 4UF, UK.

YOUNG, J. F. 1967 Humidity control in the laboratory using salt solutions—a review. *Journal of Applied Chemistry* **17,** 241–245.

The Role of Proline and Other Amino Acids in Osmoregulation of Escherichia coli B/r/1

G. D. Anagnostopoulos and G. Dhavises*

*Department of Microbiology, Queen Elizabeth College,
University of London, London, UK*

When bacteria are transferred from a normal medium to one made hypertonic by the addition of a non-penetrating solute, such as sucrose, they undergo an abrupt osmotic dehydration and the protoplast shrinks. These effects modify the physiological functions of the cytoplasmic membrane (Christian 1955b; Henneman & Umbreit 1964; Okrend & Doetsch 1969). Subsequent survival, rate of growth and ultimate population density will depend upon the water activity (a_w) of the new medium and the efficiency of the osmoregulation mechanism of the organism, in terms of rate and extent of deplasmolysis (Dhavises & Anagnostopoulos 1979a, b). Amino acids, and in particular proline, were found to be indispensable for osmotic adaptation. Proline and other amino acids are also known to be essential for the stimulation of growth and respiration at reduced a_w (Christian 1955a; Christian & Waltho 1966; Christian & Hall 1972).

Dhavises & Anagnostopoulos (1979a, b) found that deplasmolysis was an absolute requirement for resumption of growth of *E. coli* B/r/1 in a hypertonic sucrose-salts medium. It was found to be a process that could not occur in absence of cell activity. These aspects and the "hyperturgid" state of the cells, as the result of advanced deplasmolysis in the presence of proline and other amino acids, are examined further in relation to the onset and rate of subsequent growth in a hypertonic medium.

Materials and Methods

Organism and conditions of culture

Escherichia coli B/r/1 (NCIB 10772) was grown in DMA, a glucose salts medium (Pirt 1967) in shaken flasks at 37°C. Amino acid supplemented

*Present address: Biology Department, Kasetsart University, Bangkok 9, Thailand.

medium contained specified amounts of proline or 40 mg l^{-1} of proline and/or each of the following 8 amino acids: arginine, aspartic acid, glutamic acid, histidine, isoleucine, methionine, phenylalanine, valine. The inoculum was from a stationary culture diluted three-fold in sodium-potassium phosphate buffer, pH 7·2, to prevent K^+ losses during its standardization, and sedimented in a microangle centrifuge. The pellet provided an inoculum of about 2×10^9 cells ml^{-1} of the plasmolysing system.

Water activity (a_w)

The medium (a_w 0·999) was adjusted to a_w values 0·986, 0·981, 0·975, 0·960 and 0·948 by addition of sucrose (g l^{-1}) as follows, respectively: 0·635 M (217·39), 0·811 M (277·78), 0·996 M (340·09), 1·375 M (470·58) and 1·603 M (548·78). The a_w value was checked using a dew-point apparatus (Anagnostopoulos 1973).

Cell counts

Viable counts were obtained by plating, on nutrient agar, six 20 μl drops of dilutions in isotonic sodium-potassium phosphate buffer, pH 7·2, and incubating at 37°C. The specific growth rate and the definition of lag were based on a least squares analysis of the data and extrapolation. Total cell counts were obtained using a Coulter Counter model Z_{Bl} (Coulter Electronics, Dunstable, England) using a 30 μm orifice probe.

Deplasmolysis

The kinetics of the deplasmolysis process were established by optical density (D) readings at 540 nm, using a Spectronic 20 (Bauch and Lombe, Allied Research Laboratories, UK). The cell suspension was maintained at 37°C and D readings were taken at specified intervals 3 min after the mixing of inoculum. The interpretation of the data was based on suggestions made by Mitchell & Moyle (1956), Avi-Dor et al. (1956), Bovell et al. (1963) and Knowles (1971). A D increase was due to the shrinkage of E. coli, as the result of plasmolysis, and a decrease was due to the reexpansion of the protoplast as the result of rehydration.

Results

The minimum a_w for growth of E. coli B/r/1 in sucrose-DMA medium was established by experiments such as that of Fig. 1. It can be seen that the minimum a_w at which growth occurred was 0·981; below this value, which is also the minimum for deplasmolysis, conditions were

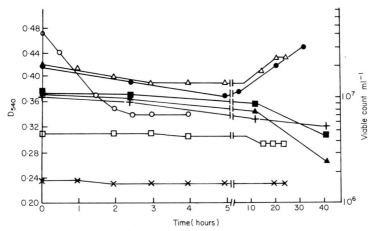

Fig. 1. Deplasmolysis and viability of *E. coli* B/r/1 in sucrose-DMA medium at 37°C. *D* (optical density) values (open symbols) and viable counts (closed symbols): a_w 0·981, ○ ●; a_w 0·975, △ ▲; a_w 0·960, □ ■; a_w 0·948, × +.

slowly lethal. In the latter case, the cells remained plasmolysed and no penetration of sucrose occurred even on prolonged incubation.

Figure 2 shows the influence of proline on deplasmolysis and subsequent growth. Proline enabled both processes to occur at a lower a_w, i.e. 0·975 but not at a_w 0·960. No deplasmolysis occurred in aqueous

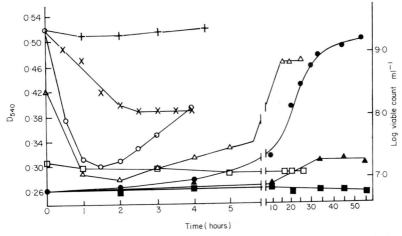

Fig. 2. Influence of proline (40 mg l^{-1}) on deplasmolysis and growth of *E. coli* B/r/1 at 37°C. Aqueous sucrose solution at a_w 0·981, *D*, + ; sucrose-DMA medium at a_w 0·981, *D*, × ; *D* (optical density) values (open symbols) and viable counts (closed symbols) in proline-sucrose-DMA medium: a_w 0·981, ○ ●; a_w 0·975, △ ▲; a_w 0·960, ○ ■.

sucrose solution. Deplasmolysis was slower in the unsupplemented sucrose-DMA medium and proceeded to the ultimate osmotic equilibrium at which growth resumed, although after a long lag (Table 1). It can be seen also that in the proline-sucrose-DMA systems at a_w 0·981

TABLE 1

Influence of amino acids on the growth of E. coli *B/r/1 in sucrose-DMA at* a_w *0·981, at 37°C*

Amino acid (mg l⁻¹)	Growth lag (h)	Specific growth rate*
Proline absent	12 or longer	0·072
Proline 10	6·4	0·124
Proline 80	5·0	0·115
Proline 240	5·0	0·119
Mixture of 8 amino acids† (40 mg l⁻¹ each)	5·4	0·130
Proline and above amino acids (40 mg l⁻¹ each)	0	0·265

*Divisions h⁻¹; mean of 2 or 3 experiments.
†Arginine, aspartic acid, glutamic acid, histidine, isoleucine, methionine, phenylalanine, valine.

and 0·975, the initial deplasmolysis was reversed towards the osmotic equilibrium observed in the sucrose-DMA medium. This reversion occurred during the lag phase of the culture (Table 1; Dhavises & Anagnostopoulos 1979*b*) and cannot be confused with the effect on *D* of the subsequent growth. The onset and rate of reversion was dependent upon the a_w of the medium (Fig. 2) and the concentration of proline (Fig. 3).

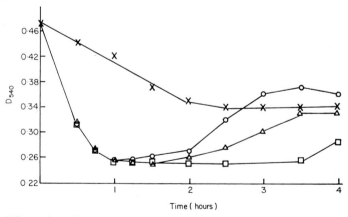

Fig. 3. Effect of proline concentration on the deplasmolysis of *E. coli* B/r/1 in sucrose-DMA medium at a_w 0·981, at 37°C. No proline added, ×; proline 10 mg l⁻¹, ○; proline 80 mg l⁻¹, △; proline 240 mg l⁻¹, □.

In the experiment of Fig. 3, the total count estimated by the Coulter Counter was constant, i.e. the change in D could not be ascribed to cell division. The increase in D observed was, therefore, due to partial reversion of deplasmolysis towards the osmotic equilibrium attained in the sucrose-DMA system. It can be seen also that the concentration of proline affected the onset and rate of the reversion process rather than the initial rate and extent of deplasmolysis.

Proline affected the growth kinetics of *E. coli* B/r/1 by reducing the lag phase and increasing the specific growth rate (Table 1). However, neither of these effects was concentration dependent above 10 mg 1^{-1} of proline. A comparable effect was shown by 8 amino acids that had been found, among 18 amino acids tested, to stimulate growth of *E. coli* in sucrose-DMA (Dhavises & Anagnostopoulos 1979a). However, when full supplementation of the sucrose-DMA medium was made with proline and the above 8 amino acids, the lag phase was abolished and the specific growth rate was doubled. Under these conditions deplasmolysis also was fastest and most extensive.

Discussion

One consequence of plasmolysis is the shrinkage of the plasma membrane and the impairment of its functions, notably of respiration (Okrend & Doetsch 1969; Henneman & Umbreit 1964). When cells resume growth in a hypertonic medium, an osmotic equilibrium with the suspending medium has been established. However, it has been shown that when proline is present, deplasmolysis proceeds beyond the ultimate osmotic equilibrium and then is reversed towards that equilibrium. This might arise as the result of proline accumulation, its partial conversion to glutamate and of the "K$^+$ influx/glutamate synthesis" two-way stimulation process (Measures 1975; Brown 1976).

The results presented here exclude sucrose permeation as a cause of deplasmolysis. In fact, the strain of *E. coli* used could not utilize sucrose as the sole carbon source in DMA and plasmolysis in aqueous sucrose solution was found to be irreversible.

The role of proline in stimulating respiration and subsequent growth of *Salmonella oranienburg* was demonstrated by Christian & Waltho (1966) and Christian & Hall (1972). It seems, however, that proline promotes also the repair of the membrane functional injuries caused by osmotic dehydration. Proline was found more effective than other amino acids used together in terms of the rate and extent of deplasmolysis (Dhavises & Anagnostopoulos 1979b), resumption of growth and growth rate. However, the presence of these amino acids enhanced further this property of proline.

Quantitative studies of the effect of proline concentration did not show a strong relationship with the growth lag, growth rate and rate and extent of deplasmolysis. This suggests a state of saturation, maybe in proline uptake above 10 mg l^{-1}. This may be also the reason for the relationship observed between proline concentration and the onset and rate of the deplasmolysis reverse process.

In conclusion, it appears that the rate and extent of the initial deplasmolysis relates closely to the onset and rate of growth in a hypertonic medium. Deplasmolysis increased in the sucrose-salts medium in the following order of supplementation: no addition, glutamate, mixture of 8 amino acids including glutamate, proline, proline and the mixture of amino acids (Dhavises & Anagnostopoulos 1979b).

References

ANAGNOSTOPOULOS, G. D. 1973 Water activity in biological systems: a dew-point method for its determination. *Journal of General Microbiology* **77**, 233–235.

AVI-DOR, Y., KUCZYNSKI, M., SCHATZBERG, G. & MAGER, J. 1956 Turbidity changes in bacterial suspensions: Kinetics and relation to metabolic state. *Journal of General Microbiology* **14**, 76–83.

BOVELL, C. R., PACKER, L. & HELGERSON, R. 1963 Permeability of *Escherichia coli* to organic compounds and inorganic salts measured by light-scattering. *Biochemical and Biophysical Acta* **75**, 257–266.

BROWN, A. D. 1976 Microbial water stress. *Bacteriological Reviews* **40**, 803–846.

CHRISTIAN, J. H. B. 1955a The influence of nutrition on the water relations of *Salmonella oranienburg*. *Australian Journal of Biological Sciences* **8**, 75–82.

CHRISTIAN, J. H. B. 1955b The water relations of growth and respiration of *Salmonella oranienburg* at 30°C. *Australian Journal of Biological Sciences* **8**, 490–497.

CHRISTIAN, J. H. B. & HALL, J. M. 1972 Water relations of *Salmonella oranienburg*: accumulation of potassium and amino acids during respiration. *Journal of General Microbiology* **70**, 497–506.

CHRISTIAN, J. H. B. & WALTHO, J. A. 1966 Water relations of *Salmonella oranienburg*: stimulation of respiration by amino acids. *Journal of General Microbiology* **43**, 345–355.

DHAVISES, G. & ANAGNOSTOPOULOS, G. D. 1979a Influence of amino acids and water activity on the growth of *Escherichia coli* B/r/1. *Microbios Letters* **7**, 105–115.

DHAVISES, G. & ANAGNOSTOPOULOS, G. D. 1979b Influence of amino acids on the deplasmolysis of *Escherichia coli* B/r/1. *Microbios Letters* **7**, 149–159.

HENNEMAN, D. H. & UMBREIT, W. W. 1964 Influence of the physical state of the bacterial cell membrane upon the rate of respiration. *Journal of Bacteriology* **87**, 1274–1280.

KNOWLES, C. J. 1971 Salt-induced changes of turbidity and volume of *Escherichia coli*. *Nature, London* **229**, 154–155.

MEASURES, J. C. 1975 Role of amino acids in osmoregulation of non-halophilic bacteria. *Nature, London* **257,** 398–400.

MITCHELL, P. & MOYLE, J. 1956 Osmotic structure and function in bacteria. *Symposium of the Society for General Microbiology* **6,** 150–180.

OKREND, A. G. & DOETSCH, R. N. 1969 Plasmolysis and bacterial motility: A method for the study of membrane function. *Archives of Microbiology* **69,** 69–78.

PIRT, S. J. 1967 A kinetic study of the mode of growth of surface colonies of bacteria and fungi. *Journal of General Microbiology* **47,** 181–197.

Effect of Sucrose and Glycerol on the Phenol Concentration Coefficient

R. G. KROLL* AND G. D. ANAGNOSTOPOULOS

*Department of Microbiology, Queen Elizabeth College,
University of London, London, UK*

Water activity (a_w) as a parameter influencing microbial growth and its effect when combined with other growth-controlling factors, such as reduced pH or temperature, has received much attention. In particular, a_w is recognized to be an important determining factor in biodeterioration and more specifically in the microbial stability of foodstuffs (Scott 1957; Corry 1973; Leistner & Rodel 1976; Brown 1976). The survival of micro-organisms after sub-lethal stress may also be modified by a_w. Stress, such as that caused by exposure to heat (Fay 1934; Baird-Parker *et al.* 1970; Corry 1974; Horner & Anagnostopoulos 1975), ethylene oxide (Phillips 1961) and radiation (Tallentire *et al.* 1963) and its relationship to a_w have been documented.

Water activity is an absolute thermodynamic concept (Scott 1957) but, when lowered by increasing concentrations of different solutes, the interaction between the particular solutes and the microbial cells can vary widely. For example, electrolytes generally produce different effects from non-electrolytes (Brown 1964). Solutes can also be divided into those which penetrate the cell and those which do not. However, the exact permeability characteristics differ according to the organism, the selectivity of the cytoplasmic membrane and, in Gram negative bacteria, on the ability to penetrate the outer membrane. The permeability of the cytoplasmic membrane depends largely on the transport systems present. Thus, non-penetrating solutes, such as sucrose, sodium chloride or calcium chloride, cause plasmolysis (Scheie 1969) while penetrating solutes, such as glycerol or urea, cause a short-lasting plasmolysis followed by deplasmolysis (Alemohammad & Knowles 1974). The stability and integrity of the membrane is affected by both types of stress (Rose 1976) and might, therefore, affect survival.

Stretching or contraction of the cytoplasmic membrane affects the

* Present address: Department of Microbiology, University of Aberdeen, Marischal College, Aberdeen, UK.

membrane pore size (Marquis & Corner 1967; Burton 1970; MacLeod *et al.* 1978) and hence the permeability to different agents. Moreover, the presence of membrane-active agents, such as phenol, might alter the characteristics of the membrane and thus the penetration characteristics of different solutes.

These aspects acquire particular importance in antimicrobial chemical preservation, where the osmotic state of the microbial cell may affect the uptake and activity of an antimicrobial agent. Antimicrobial agents are widely used in the food, cosmetic and pharmaceutical industries as preservatives. As legislation becomes progressively more restrictive in the use of preservatives, data on factors influencing their efficiency are of profound importance, besides reflecting basic biological phenomena.

We have noticed that reducing the a_w of phenol solutions affects drastically the concentration coefficient of phenol (Anagnostopoulos, unpublished data). This report concerns phenol, a typical membrane active agent (Hamilton 1968) and the kinetics of its action in relation to the a_w and the type of solute of the microbial suspension.

Materials and Methods

Preparation of cell suspensions

Serratia marcescens B8 (QEC) and *Pseudomonas aeruginosa* (NCTC 6749) were grown in nutrient broth at 30° and 37°C respectively, for 18 h. The cultures were centrifuged at 5000 g for 10 min, washed, resuspended and diluted to $2-4 \times 10^7$ cells ml^{-1} in phosphate buffer (PB: 0.1 M, pH 7.0). The diluted suspensions were distributed in 1 ml volumes in Bijou bottles and stored for the day at 4°C until 30 min before an experiment when they were placed at 25°C.

Preparation of phenol-solute mixtures

These were prepared by adding 5 ml quantities of sucrose (LR) or glycerol (LR) at twice the final required strength in PB to appropriate volumes of phenol (5 % w/v; AR) in PB. The volumes were adjusted to 9 ml with PB. (No account of PB was taken in calculating the final a_w because its influence on a_w was so small.)

Determination of bactericidal action

Throughout the reaction the mixture was stirred at 25°C. The reaction was initiated by the addition of 1 ml of cell suspension. Samples were taken at appropriate time intervals and immediately diluted in a solution, isotonic to the same plus 1% peptone (Oxoid) in PB, followed by quintuplicate 20 μl drop surface inoculation of well dried nutrient agar plates. Sampling intervals were arranged so that death was followed over an approximately two log inactivation. The plates were then incubated at 30°C until visible colonies appeared (18–36 h). Six such viable counts were made for each phenol-solute combination.

Treatment of data

Decimal reduction (D) time values for each phenol-solute concentration were determined by a least squares analysis. The concentration co-efficient "n" of phenol was similarly determined using six D value/phenol concentration sets of data and applying the relationship $k=C^n t$ (Watson 1908), where k=constant, C=phenol concentration (% w/v) and t=time for a fixed percentage lethality.

Results

The lethal effects of aqueous solutions of sucrose and glycerol were followed by experiments lasting 600 h for *P. aeruginosa* and 80 h for *S. marcescens*. The D values, some of them obtained by extrapolation, appear in Table 1. Lethality is clearly insignificant, but, for other solutes, e.g. urea, sodium chloride, lethality may be considerable.

TABLE 1

Lethality attributable to the aqueous solutions of sucrose and glycerol

	D value in hours at 25°C		
	a_w 0·980	a_w 0·950	a_w 0·920
S. marcescens			
Sucrose	194·2	104·7	55·96
Glycerol	197·3	124·4	89·96
P. aeruginosa			
Sucrose	10 030	995	293·2
Glycerol	infinite	5780	656·1

The lethality rates in phenol at concentrations ranging from 0·4 to 1·2% (w/v) and at a_w values ranging from 1·00 in PB alone to 0·920 in the presence of sucrose or glycerol, were derived and treated as described in the methods. The results appear in Table 2 and graphically in Figs 1–4.

TABLE 2

Effect of reduced a_w on the concentration coefficient 'n' of phenol

a_w	c^*	$r\dagger$	$n\ddagger$
S. marcescens—glycerol			
1·00	−0·308	−0·988	6·95
0·980	−0·328	−0·982	6·92
0·950	−0·476	−0·997	6·23
0·920	−0·299	−0·996	5·47
S. marcescens—sucrose			
1·00	−0·308	−0·988	6·95
0·980	−0·047	−0·982	6·78
0·950	−0·020	−0·984	7·67
0·920	−0·511	−0·982	9·24
P. aeruginosa—glycerol			
1·00	−0·328	−0·986	6·33
0·975	0·021	−0·995	5·32
0·950	0·561	−0·981	4·93
0·925	1·251	−0·985	4·64
P. aeruginosa—sucrose			
1·00	−0·328	−0·986	6·33
0·975	−0·478	−0·987	6·28
0·950	−0·428	−0·990	6·75
0·925	−1·422	−0·997	8·95

*Intercept of the log D—log phenol concentration straight line.
†Correlation coefficient.
‡Concentration coefficient of phenol.

Table 2 shows that the concentration coefficient of phenol, which, at a_w 1·00, was 6·33 and 6·95 for *P. aeruginosa* and *S. marcescens* respectively, was reduced by glycerol to 4·64 and 5·47 and increased by sucrose to 8·95 and 9·24 when the a_w was reduced to 0·925 and 0·920 respectively.

Figures 1–4 show the effects of solute and a_w value on the D values for the two organisms. Basically the pattern is comparable in both cases, the D value increasing with the concentration of glycerol towards a_w 0·920 and decreasing with that of sucrose. This trend is especially noticeable with *P. aeruginosa* where the D values at a_w 0·925 are as much as 50-fold higher in the presence of glycerol than at a_w 1·00. With both organisms, the sucrose-system exhibited D values that decreased sharply at a_w 0·920 and 0·925 when the phenol concentration exceeded 0·5 %.

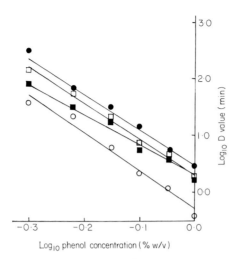

Fig. 1. Logarithmic plot of the D values for *S. marcescens* against phenol concentration in PB and glycerol solutions (○, a_w 1·00 (water); □, a_w 0·98; ●, a_w 0·95; ■, a_w 0·92).

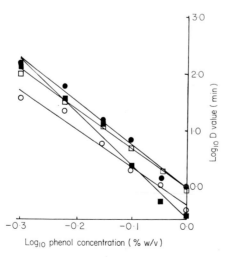

Fig. 2. Logarithmic plot of the D values for *S. marcescens* against phenol concentration in PB and sucrose solutions (for key see Fig. 1).

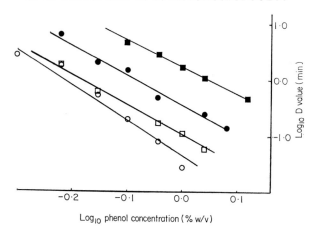

Fig. 3. Logarithmic plot of the D values for *P. aeruginosa* against phenol concentration in PB and glycerol solutions (for key see Fig. 1).

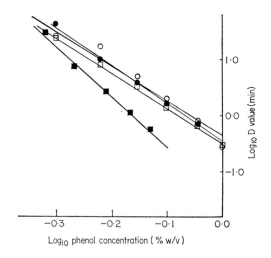

Fig. 4. Logarithmic plot of the D values for *P. aeruginosa* against phenol concentration in PB and sucrose solutions (for key see Fig. 1).

Discussion

The mode of action of phenol is multifacious but its main target site at low concentrations is the cytoplasmic membrane (Hamilton 1968). Sucrose and glycerol reduce the a_w of an aqueous cell suspension but the two solutes exert different cytological effects. Sucrose, being non-penetrant, causes plasmolysis (Scheie 1969; Scheie & Rehberg 1972).

Plasmolysis is a non-reversible phenomenon in metabolically inactive cells while they remain suspended in the sucrose solution (Dhavises & Anagnostopoulos, unpublished data). Glycerol, being a cell penetrant, causes an initial short-lasting plasmolysis followed by deplasmolysis as glycerol enters the cell (Alemohammad & Knowles 1974). It might be expected that either of these types of membrane stress would increase the bactericidal activity of phenol above that of aqueous solutions. However, the results presented here show that this was rarely the case, the bactericidal activity of phenol being reduced drastically except at a_w 0·920 and 0·925 with sucrose. This represents a state of extreme plasmolysis (Scheie 1969) where the increased activity of phenol could be related to the contracted state of the plasma membrane. This might cause greater exposure of the membrane to the phenol contained in the periplasmic space of the cell.

On the other hand, the different cytological effects of glycerol and sucrose might be responsible for their opposing influences on the concentration coefficient of phenol. As shown in Table 2, the concentration coefficient of phenol decreases as the concentration of glycerol increases towards a_w 0·920 and the reverse appears to be the case with sucrose. The significance of this effect is underlined by the exponential function of "n" in the equation $k = C^n t$ and by the corresponding magnitude of change in disinfection time in relation to changes in the concentration of phenol. These findings may have particular significance in systems such as food, cosmetics and pharmaceuticals, where a low initial dose of preservative may have its concentration reduced by interaction, absorption etc. Conversely, a low initial dose of preservative may have its activity enhanced if the solute used mimics glycerol in its action.

Besides implications for the practical preservation of biodegradable materials, these findings might lend themselves to a further understanding of the osmotic injury of the plasma membrane. Lastly, apart from the distinctly different cytological effect of the two solutes, glycerol is known to be a solvent of phenol with a high affinity for it (Kostenbauder 1968). This property might then affect the rate of phenol uptake by the cells and, consequently, the kinetics of its action as well as its concentration coefficient, thereby exhibiting the role of a "slow-release" agent for phenol.

The authors thank the Science Research Council and Boots Company Ltd. for their grants to R. G. Kroll.

156   R. G. KROLL AND G. D. ANAGNOSTOPOULOS

References

ALEMOHAMMAD, M. M. & KNOWLES, C. J. 1974 Osmotically induced volume and turbidity changes of *Escherichia coli* to salts, sucrose and glycerol, with particular reference to the rapid permeation of glycerol into the cell. *Journal of General Microbiology* **82**, 125–142.

BAIRD-PARKER, A. C., BOOTHROYD, M. & JONES, E. 1970 Effect of water activity on the heat resistance of heat sensitive and heat resistant strains of *Salmonella*. *Journal of Applied Bacteriology* **33**, 515–522.

BROWN, A. D. 1964 Aspects of bacterial response to the ionic environment. *Bacteriological Reviews* **28**, 296–329.

BROWN, A. D. 1976 Microbial water stress. *Bacteriological Reviews* **40**, 803–846.

BURTON, A. C. 1970 The stretching of pores in a membrane. In *Permeability and Function of Biological Membranes* eds Katchalskey, A. *et al*. pp. 1–19, Amsterdam: North-Holland Publishing Co.

CORRY, J. E. L. 1973 The water relations and heat resistance of micro-organisms. *Progress in Industrial Microbiology* **12**, 73–108.

CORRY, J. E. L. 1974 The effect of sugars and polyols on the heat resistance of *Salmonella*. *Journal of Applied Bacteriology* **37**, 31–43.

FAY, A. C. 1934 The effect of hypertonic sugar solutions on the thermal resistance of bacteria. *Journal of Agricultural Research* (Washington) **48**, 453–468.

HAMILTON, W. A. 1968 The mechanism of the bacteriostatic action of tetrachlorosalicylamide. *Journal of General Microbiology* **50**, 441–458.

HORNER, K. J. & ANAGNOSTOPOULOS, G. D. 1975 Effect of water activity on the heat survival of *Staphylococcus aureus, Salmonella typhimurium* and *Salm. senftenberg*. *Journal of Applied Bacteriology* **38**, 9–11.

KOSTENBAUDER, H. B. 1968 Physical factors influencing the activity of antimicrobial agents. In *Disinfection, Sterilization and Preservation* eds Lawrence, C. A. & Block, S. S. p. 45, Philadelphia: Lea and Febiger.

LEISTNER, L. & RODEL, W. 1976 Inhibition of micro-organisms by water activity. In *Inhibition and Inactivation of Vegetative Microbes, Society for Applied Bacteriology* Symposium No. 5, eds Skinner, F. A. & Hugo, W. B., pp. 219–237, London: Academic Press.

MACLEOD, R. A., GOODBODY, M. & THOMSON, J. 1978 Osmotic effects on membrane permeability in a marine bacterium. *Journal of Bacteriology* **133**, 1135–1143.

MARQUIS, R. E. & CORNER, T. R. 1967 Permeability changes associated with osmotic swelling of bacterial protoplasts. *Journal of Bacteriology* **93**, 1177–1178.

PHILLIPS, C. R. 1961 The sterilizing properties of ethylene oxide. In *Recent Developments in the Sterilization of Surgical Materials*, pp. 59–75, London: Pharmaceutical Press.

ROSE, A. H. 1976 Osmotic stress and microbial survival. In *The Survival of Vegetative Microbes, Society for General Microbiology* Symposium No. 26, eds Gray, T. R. G. & Postgate, J. R., pp. 155–182, Cambridge University Press.

SCHEIE, P. O. 1969 Plasmolysis of *Escherichia coli* B/r with sucrose. *Journal of Bacteriology* **98**, 335–340.

SCHEIE, P. O. & REHBERG, R. 1972 Response of *Escherichia coli* B/r to high concentrations of sucrose in a nutrient medium. *Journal of Bacteriology* **109**, 229–235.

SCOTT, W. J. 1957 Water relations of food micro-organisms. *Advances in Food Research* **7**, 83–127.

TALLENTIRE, A., DICKINSON, N. A. & COLLET, J. M. 1963 A dependence on water content of bactericidal efficiency of gamma radiation. *Journal of Pharmacy and Pharmacology* supp. **15**, 180T–181T.

WATSON, H. E. 1908 A note on the variation of the rate of disinfection with change in concentration of disinfectant. *Journal of Hygiene, Cambridge* **8**, 536–542.

The Preparation of Bacterial Spores for Evaluation of the Sporicidal Activity of Chemicals

W. M. WAITES AND CATHERINE E. BAYLISS

Agricultural Research Council, Food Research Institute, Norwich, UK

Numerous methods have been devised to examine the destruction of spores by chemicals. Such tests usually measure either the rate of kill (Mossel 1963; Sykes 1970; Forsyth, 1975) or the time taken for complete destruction (Horwitz 1970; Stark *et al.* 1975). Whichever method is used, the estimate of sporicidal efficiency of chemicals ultimately depends on the resistance of the spores tested. Detailed studies have been made of methods of preparing different batches of spores of the same organism with the same heat resistance (Friesen & Anderson 1974; Heintz *et al.* 1976) but less effort has gone into reproducible production of spores with a consistent resistance to chemicals. Many of the factors which influence heat resistance also influence chemical resistance (for reviews, see Roberts & Hitchins 1969; Russell 1971) but they may act in different ways, so that spores with extreme heat resistance may not be particularly resistant to chemicals. In addition, the order of heat resistance of the most economically important spores is known so that a logical choice can be made to test the efficiency of heat processing. For example, in the heat processing of foods, the spores used in trials with inoculated packs have either an extremely high heat resistance so that their complete destruction will produce sterility (spores of *Bacillus stearothermophilus*) or a lower but still high heat resistance which is slightly greater than that of *Clostridium botulinum*, the most important mesophilic spore-former (e.g. spores of *Cl. sporogenes* PA3679; National Canners Association 1968). The choice of strains to produce spores to test the efficiency of chemical sporicides is much less obvious so that the strains chosen should include those on which the chemical must act in practice.

In general, before testing the destruction of spores by chemicals, the following considerations are important: (1) organisms; (2) sporulation media; (3) harvesting and washing procedure; (4) storage conditions; (5) state of spores at time of test; (6) recovery conditions.

Organisms

Compared to vegetative cells, bacterial spores may be up to 10 000 times as resistant to chemicals (Phillips 1952) although resistance may vary markedly between species and strains (Fig. 1). The relative resistance of spores to different chemicals may also differ with the strain. For example,

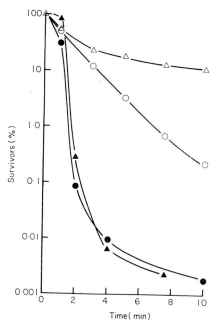

Fig. 1. Destruction of spores of different strains of *B. subtilis* by hydrogen peroxide. Spores of *B. subtilis* SA22 (○), B17 (●), 713 (△) and 706 (▲), produced on bacillus spore medium agar (Franklin *et al.* 1970) at 30°C (B17) or 37°C, were incubated at 90°C with 0·5% (w/v) hydrogen peroxide. Samples (1 ml) were removed at intervals, diluted and plated after incubation with catalase (7650 u) (Sigma Ltd). 706 and 713 were plated on bacillus spore medium and B17 and SA22 on plate count agar (Oxoid).

dried spores of *Cl. sporogenes* have a higher resistance to ethylene oxide (Beloian 1977) but lower resistance to phenol than those of *B. subtilis* var. *niger* (Sykes 1965). It is clear that it is not possible to extrapolate from the sporicidal efficiency of a particular chemical with one strain to other chemicals or strains. In addition, heat resistance has been shown to be unrelated to chemical resistance at least for phenol (Briggs 1966), ethylene oxide (El-Bisi *et al.* 1963), chlorine (Dye & Mead 1972) and hydrogen peroxide (Toledo *et al.* 1973). Hence, spores with extreme resistance to wet or dry heat cannot necessarily be used to determine the sporicidal efficiency of chemicals. The relative resistance of spores of different strains to chemicals at low temperature may not be the same as at high temperatures (Toledo 1975) so that rates of destruction cannot be extrapolated from low to high temperatures.

In general, determination of sporicidal efficiency with spores of only one bacterial species may not provide an accurate index to sterilizing efficiency (Beloian 1977). Strains should be chosen which are relevant to the use of sporicide (Borich & Pepper 1970), together with those producing spores of known high resistance.

Sporulation Media

Many of the points discussed under the previous heading are relevant in choosing the sporulation medium. Partly because of the problems of removing vegetative cells, sporulation media are designed to produce a high percentage of free spores with few vegetative cells or germinated spores, but there has been no comparison of the resistances of spores from such media and those formed on media which produce only a few free spores. Spores produced on some media may, for example, have little resistance to chemicals but a relatively high heat resistance. Further, spores grown on different batches of the same medium may differ in their resistance so that each batch should be examined to make certain that the spores do have the expected resistance. The use of 2·5 M HCl as a standard to compare resistance of different spores has been suggested (Ortenzio *et al.* 1953) but resistance to other chemicals does not parallel that against HCl (Beloian 1977).

Most strains of *B. cereus*, *B. coagulans*, *B. megaterium*, *B. licheniformis* and *B. subtilis* will sporulate well at 30°C on a potato based medium (Gould & Ordal 1968) containing (g l^{-1}): potato extract (Difco), 4; yeast extract (Difco), 4; glucose, 2·5; agar, 15; adjusted to pH 7·2.

Some proteolytic clostridia will sporulate in a medium (Brown *et al.* 1957) containing (g l^{-1}): trypticase (BBL), 30; (NH$_4$)$_2$SO$_4$, 10; sodium thioglycollate, 1; adjusted to pH 7·2.

Certain saccharolytic clostridia sporulate on media based on trypticase and glucose (Bergère & Hermier 1965) or potato (Lund *et al.* 1978). In general, clostridial spores are more difficult to produce than those of *Bacillus* spp. but media for particular strains have been discussed (Gould 1971). A higher percentage of spores may be formed on solid rather than liquid media and even clostridial spores will form on agar in Petri dishes incubated anaerobically (Waites & Wyatt 1971; Waites *et al.* 1979) although spores produced on agar media may show greater variation than those formed in liquid media. Small additions to a medium may markedly change resistance, for example, addition of glucose and metal salts to nutrient agar (Setlow & Kornberg 1969) increased the resistance of spores of *B. subtilis* to hydrogen peroxide (Fig. 2).

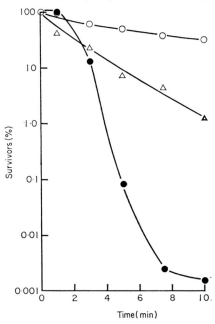

Fig. 2. Effect of sporulation media on destruction of spores of *B. subtilis* SA22 by hydrogen peroxide. Spores were produced on supplemented nutrient medium agar (Setlow & Kornberg 1969) (○), bacillus spore medium agar (Franklin *et al.* 1970) (△) or nutrient broth (Oxoid) plus New Zealand agar (1·2% w/v) (●) before treating with H_2O_2 and plating as described for Fig. 1.

A chemically defined medium has been reported to produce spores of *B. subtilis* which are less variable in resistance to glutaraldehyde than those produced on a soil-extract medium (Stark *et al.* 1975; Forsyth 1975). In addition, the age of the spores on harvesting may be important since the resistance of spores of *B. subtilis* to glutaraldehyde reached a maximum after 7 days and then declined rapidly although complete sporulation had occurred after 3 days (Forsyth 1975). It is apparent that, if possible, the resistance of spores produced on several different media should be compared before selection of the medium producing the highest percentage of the most resistant spores.

Harvesting and Washing

In natural environments spores are unwashed and are likely to be 'dirty'. It has been suggested that unwashed spores should be used for resistance studies. However, spores produced in the laboratory have usually been formed in rich media so that washing may be required to remove germinants, especially if the spores are to be stored. 'Soil' as organic matter may be added at the time of test (see below).

Liquid media containing spores should be harvested by centrifugation after cooling to 4°C while spores produced on solid media can be scraped from the surface with a polypropylene spatula in the presence of cold, sterile glass-distilled water. Spores should be washed at 4°C with a minimum of 50 ml (for each wash) of sterile glass-distilled water for 10^{11} spores and at least five washes (Waites & Wyatt 1971). High centrifugation speeds may germinate spores of some strains so that the minimum centrifugation speed required to pellet the spores should be used.

Incubation with enzymes such as trypsin (100 μg ml^{-1}; Grecz et al. 1962), papain (1 mg ml^{-1}; Warth et al. 1963), lysozyme (200 μg ml^{-1}; Brown et al. 1957) or deoxyribonuclease (Murrell 1969) have been used to remove debris left by lysed sporangia or vegetative cells. Ultrasonication (Grecz et al. 1962; St Julian et al. 1967; Burgos et al. 1972) has also been used to break sporangia. However, old cells may sometimes be resistant to lysozyme (Forsyth 1975) and all such treatments should be considered carefully as they may decrease the resistance of the spores (St Julian et al. 1967). In particular, centrifugation on density gradients (Tamir & Gilvarg 1966) and the two phase system of Sacks & Alderton (1961) have been used to separate free spores from germinated spores and vegetative cells, but may initiate germination of a fraction of the spore population of some strains (Murrell & Warth 1965; Murrell 1969). Electrophoresis might also be used to separate dormant spores from germinated spores and vegetative cells (Douglas 1959). Washing by centrifugation with 0·03 м HCl has been used to remove metal precipitates (Slepecky & Foster 1959) but may also remove metal ions from the spores themselves (Waites et al. 1979). Germinated spores or debris from vegetative cells may neutralize particularly reactive chemicals such as chlorine more rapidly than the spores themselves so that the effective concentration of the chemical is decreased.

In conclusion, one method of harvesting and washing cannot be used for all strains or even for one strain grown on several different media. If possible, the medium used should allow production of at least 90 % free spores so that post-harvest manipulation can be reduced to a minimum.

Storage Conditions

Resistance of spores may change on storage so that destruction immediately after harvesting should be compared with that after storing for at least 3 months. For example, decreasing resistance to glutaraldehyde during storage at 4°C and increasing resistance to HCl during storage *in vacuo* after lyophilization were observed by Forsyth (1975) while Bomar (1962) found that resistance to ethylene oxide also changed during storage. Wallen (1976) reported that spores stored in the presence of NaCl retained resistance to H_2O_2 longer than those stored in water alone, but the presence of NaCl in the recovery medium may reduce heat resistance (Roberts *et al.* 1966) so that all NaCl should be removed before recovering. Thawing and re-freezing spores stored frozen may reduce resistance and stimulate germination. The resistance of spores stored at

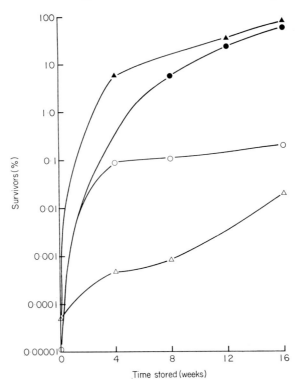

Fig. 3. Effect of storage on destruction of spores of *B. subtilis* var. *niger* by chlorine. Spores were produced on bacillus spore medium agar (Franklin *et al.* 1970) (○, ●) or potato medium agar (Gould & Ordal 1968) (△, ▲) before storing at —20°C (○, △) or +2°C (●, ▲) and treating with 100 μg ml $^{-1}$ free chlorine at 1°C for 30 min. Treated spores were diluted and plated on plate count agar (Oxoid).

+2°C may increase more rapidly than that of spores stored at —20°C (Fig. 3). Freeze drying damages spores of some strains (Marshall *et al.* 1963) and an alternative is to store in ethanol (Molin & Östlund 1976) although higher alcohols are sporicidal over long periods (Zuk & Parisi 1974). Ethanol must be removed before testing spore destruction as it inhibits germination (Waites & Wyatt 1974; Dring & Gould 1975).

State of Spores at Time of Test

Sporicidal activity should not be examined in the presence of possible germinants unless it is desired to test the resistance of germinated spores (Gould 1971). Dry spores may be more resistant to chemicals than wet spores and Stuart (1969) has suggested that spores should be vacuum dried to test the sporicidal efficiency of both gaseous and liquid chemicals. Spores of *Cl. sporogenes* dried in a vacuum over calcium chloride have much greater resistance to HCl than spores dried under atmospheric conditions (Ortenzio 1966) while spores dried from saline were more resistant to ethylene oxide than those dried from water or ethanol (Kaye & Phillips 1949; Beeby & Whitehouse 1965). In addition, spores dried from methanol were more stable during storage than those dried from water (Beeby & Whitehouse 1965). However, there are conflicting reports about the relative resistances of wet and dry spores to hydrogen peroxide (Ito *et al.* 1973; Toledo *et al.* 1973).

An increase in the number of spores present will increase the time required for sterilization (Spaulding *et al.* 1977) and a large fraction of the population of spores of some strains will not form colonies unless heat activated. For example, spores of *Cl. sporogenes*, but not *B. subtilis*, incubated at 37°C after treatment with HCl did not grow until heated at 80°C for 20 min (Ortenzio 1966). Some chemicals activate spores (furfural, Mefferd & Campbell 1951; iodophors, Cousins & Allan 1967; chlorine Wyatt & Waites 1973, hydrogen peroxide, Bayliss & Waites 1976) and treatment at pH 1·5 (Lewis *et al.* 1965) or with chlorine (Dye & Mead 1972) or hydrogen peroxide (Toledo *et al.* 1973) has been shown to reduce heat resistance. Sublethal pre-heating may decrease resistance to phenol and to formaldehyde (Reddish 1950), glutaraldehyde (Rubbo *et al.* 1967) and hydrogen peroxide (Toledo *et al.* 1973) although the resistance of spores which do not require heat activation for germination may not be reduced.

The carrier to which spores are attached may also influence their resistance. For example, Kaye & Phillips (1949) and Gilbert *et al.* (1964) found that spores dried on hard impermeable surfaces were less susceptible to ethylene oxide than those dried on permeable surfaces.

Finally, organic matter reduces sporicidal efficiency, especially of particularly reactive chemicals such as chlorine (Spaulding et al. 1977). Organic matter ('soil') can be added immediately before testing although suitable controls should be used to determine if germination, and consequent loss of resistance, is initiated. Yeast, blood and serum have been used as soil (Trueman 1971).

Ideally the destruction of both wet and dry, heat activated and unactivated spores should be compared on the surfaces to be sterilized under the conditions which will be used in practice.

Recovery Conditions

The sporicidal efficiency of a chemical is usually determined by detecting the growth of viable spores. The conditions for spore germination and outgrowth and for division of the vegetative cell must be as near optimal as possible since damaged spores are particularly susceptible to inhospitable conditions (Schmidt 1955; Roberts 1970; Futter & Richardson 1970a,b). Wherever possible a suitable neutralizing or quenching agent should be used to prevent any carry over of the chemical into the growth medium. Examples of neutralizing agents (MacKinnon 1974) which have been used are catalase (for hydrogen peroxide), sodium bisulphite (for formaldehyde and glutaraldehyde; Rubbo et al. 1967; Forsyth 1975; although glycine has also been suggested for glutaraldehyde), sodium thiosulphate (for iodine; Rubbo et al. 1967, and chlorine; Trueman 1971), lecithin-Tween-80 (for quaternary ammonium compounds; Quisno et al. 1946) and sodium sulphite (for formaldehyde; Forsyth 1975). Universal neutralizing solutions have also been suggested (Mossel 1963; Engley & Dey 1970).

For the sporicidal test of the Association of Official Analytical Chemists two formulations of thioglycollate broth are commonly used to recover treated spores (Horwitz 1970). Thioglycollate broth may not produce maximal recovery, however, since Friedl (1955) found that trypticase soy broth produced more efficient recovery of small numbers of spores of B. subtilis. Addition of thioglycollate to media decreases clostridial growth (Barker & Wolf 1971) and Mossel & Beerens (1968) have recommended cysteine-HCl as a replacement.

In general, growth media and conditions which support the highest viable count of untreated spores should be used. Damaged spores will be at least as sensitive to inhospitable conditions and may have more specialized requirements. In particular, damaged spores will lose viability if there is a delay before they are added to recovery media (Futter & Richardson 1972) and clostridial spores which have been sufficiently damaged to allow development only after a lag will require anaerobic conditions to be maintained for longer periods than undamaged spores. Viable counts can also increase after long incubation times (Futter & Richardson 1972; Forsyth 1975) especially on sub-optimal media (Fig. 4) so that long lag periods before growth occurs may provide evidence of sub-optimal recovery conditions.

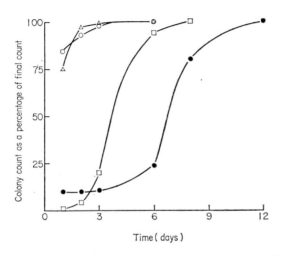

Fig. 4. Effect of treatment with chlorine on colony formation of spores of *B. subtilis* var. *niger*. Spores were produced on potato medium agar and treated with 0 (○), 25 (△), 50 (□) or 100 (●) μg ml^{-1} free chlorine at 1°C for 30 min as described by Wyatt & Waites (1975) before diluting and plating on potato medium agar (Gould & Ordal 1968). The colonies formed were counted after 1, 2, 3, 6, 8 and 12 days incubation at 30°C. Survivors (as a percentage of untreated spores) were: 25 μg ml^{-1}, 118; 50 μg ml^{-1}, 38; 100 μg ml^{-1}, 0·19.

Conclusions

In testing the sporicidal efficiency of chemicals, consideration must be given to each stage of spore preparation and destruction (Fig. 5). Until sufficient evidence of the resistance of natural spore populations to a particular chemical has accumulated, care should be taken to examine the destruction of the most resistant spores available. This can only be achieved by comparison of the destruction of spores of several different strains, produced on several different media, tested under different conditions and recovered under optimal conditions. Such requirements can only be met under ideal circumstances and results achieved in less rigorous circumstances should be treated with caution until the character-

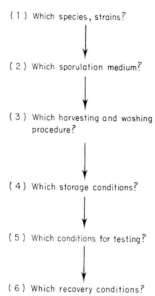

(1) Which species, strains?

(2) Which sporulation medium?

(3) Which harvesting and washing procedure?

(4) Which storage conditions?

(5) Which conditions for testing?

(6) Which recovery conditions?

Fig. 5. Considerations required before testing the sporicidal efficiency of a chemical.

istics of both the spores and the chemical under test have been more fully determined.

We thank Mrs L. R. Wyatt for the results described in Fig. 4, Mrs L. J. Wallwork and Mrs H. A. Gravenall for skilled technical assistance and Drs B. M. Lund, G. C. Mead and J. Payne for critical reading of the manuscript.

References

BARKER, A. N. & WOLF, J. 1971 Effects of thioglycollate on the germination and growth of some clostridia. In *Spore Research—1971* eds Barker, A. N., Gould, G. W. & Wolf, J. pp. 95–109, London: Academic Press.

BAYLISS, C. E. & WAITES, W. M. 1976 The effect of hydrogen peroxide on spores of *Clostridium bifermentans*. *Journal of General Microbiology* **96**, 401–407.

BEEBY, M. M. & WHITEHOUSE, C. E. 1965 A bacterial spore test piece for the control of ethylene oxide sterilization. *Journal of Applied Bacteriology* **28**, 349–360.

BELOIAN, A. 1977 Methods of testing for sterility and efficacy of sterilizers, sporicides and sterilizing processes. In *Disinfection, Sterilization and Preservation* ed. Block, S. S. pp. 11–48, Philadelphia: Lea & Febiger.

BERGÉRE, J. L. & HERMIER, J. 1965 Étude des facteurs contrôlant la croissance et la sporulation de *Clostridium butyricum*. *Annales de l'Institut Pasteur* **109**, 80–89.

BOMAR, M. 1962 The relationship between the age of *Bacillus subtilis* spores and their resistance to ethylene oxide. *Folia Microbiologica* **7**, 259–261.

BORICH, P. M. & PEPPER, R. E. 1970 *Disinfection* pp. 85–102, New York: Marcel Dekker.

BRIGGS, A. 1966 The resistance of spores of the genus *Bacillus* to phenol, heat and radiation. *Journal of Applied Bacteriology* **29**, 490–504.

BROWN, W. L., ORDAL, Z. J. & HALVORSON, H. O. 1957 Production and cleaning of spores of Putrefactive Anaerobe 3679. *Applied Microbiology* **5**, 156–159.

BURGOS, J., ORDÓÑEZ, J. A. & SALA, F. 1972 Effect of ultrasonic waves on the heat resistance of *Bacillus cereus* and *Bacillus licheniformis* spores. *Applied Microbiology* **24**, 497–498.

COUSINS, C. M. & ALLAN, C. D. 1967 Sporicidal properties of some halogens. *Journal of Applied Bacteriology* **30**, 168–174.

DOUGLAS, H. W. 1959 Electrophoretic studies on bacteria Part 5.—Interpretation of the effects of pH and ionic strength on the surface charge borne by *B. subtilis* spores, with some observations on other organisms. *Transactions of the Faraday Society* **55**, 850–856.

DRING, G. J. & GOULD, G. W. 1975 Electron transport-linked metabolism during germination of *Bacillus cereus* spores. In *Spores VI* eds Gerhardt, P., Costilow, R. N. & Sadoff, H. L. pp. 488–494, Washington: American Society for Microbiology.

DYE, M. & MEAD, G. C. 1972 The effect of chlorine on the viability of clostridial spores. *Journal of Food Technology* **7**, 173–181.

EL-BISI, H. M., VONDELL, R. M. & ESSELEN, W. B. 1963 Studies on the kinetics of the bactericidal action of ethylene oxide in the vapor phase. *Bacteriological Proceedings* p. 13.

ENGLEY, F. B. & DEY, B. P. 1970 A universal neutralizing medium for antimicrobial chemicals. *Chemical Specialties Manufacturers Association Proceedings* 100–106.

FORSYTH, M. P. 1975 A rate of kill test for measuring sporicidal properties of liquid sterilizers. *Developments in Industrial Microbiology* **16**, 37–47.

FRANKLIN, J. G., UNDERWOOD, H. M., PERKIN, A. G. & BURTON, H. 1970 Comparison of milks processed by the direct and indirect methods of ultra-high-temperature sterilization. II. The sporicidal efficiency of an experimental plant for direct and indirect processing. *Journal of Dairy Research* **37**, 219–226.

FRIEDL, J. L. 1955 Report on sporicidal tests for disinfectants. *Journal of the Association of Official Agricultural Chemists* **38**, 280–287.

FRIESEN, W. T. & ANDERSON, R. A. 1974 Effects of sporulation conditions and cation-exchange treatment on the thermal resistance of *Bacillus stearothermophilus* spores. *Canadian Journal of Pharmaceutical Sciences* **9**, 50–53.

FUTTER, B. V. & RICHARDSON, G. 1970a Viability of clostridial spores and the requirements of damaged organisms I. Method of colony count, period and temperature of incubation, and pH value of the medium. *Journal of Applied Bacteriology* **33**, 321–330.

FUTTER, B. V. & RICHARDSON, G. 1970b Viability of clostridial spores and the requirements of damaged organisms II. Gaseous environment and redox potentials. *Journal of Applied Bacteriology* **33**, 331–341.

FUTTER, B. V. & RICHARDSON, G. 1972 Viability of clostridial spores and the requirements of damaged organisms III. The effect of delay in plating after exposure to the bactericidal influence. *Journal of Applied Bacteriology* **35**, 301–307.

GILBERT, G. L., GAMBILL, V. M., SPINER, D. R., HOFFMAN, R. K. & PHILLIPS, C. R. 1964 Effect of moisture on ethylene oxide sterilization. *Applied Microbiology* **12**, 496–503.

GOULD, G. W. 1971 Methods for studying bacterial spores. In *Methods in Microbiology* Vol. 6A, eds Norris, J. R. & Ribbons, D. W. pp. 326–381, London: Academic Press.

GOULD, G. W. & ORDAL, Z. J. 1968 Activation of spores of *Bacillus cereus* by γ-radiation. *Journal of General Microbiology* **50**, 77–84.

GRECZ, N., ANELLIS, A. & SCHNEIDER, M. D. 1962 Procedure for cleaning of *Clostridium botulinum* spores. *Journal of Bacteriology* **84**, 552–558.

HEINTZ, M.-T., URBAN, S., SCHILLER, I., GAY, M. & BÜHLMANN, X. 1976 The production of spores of *Bacillus stearothermophilus* with constant resistance to heat and their use as biological indicators during the development of aqueous solutions for injection. *Pharmaceutica Acta Helvetiae* **51**, 137–143.

HORWITZ, W. 1970 *Official Methods of Analysis of the Association of Official Analytical Chemists* pp. 64–65. Washington: Association of Official Analytical Chemists.

ITO, K. A., DENNY, C. B., BROWN, C. K., YAO, M. & SEEGER, M. L. 1973 Resistance of bacterial spores to hydrogen peroxide. *Food Technology* **27**, 58–66.

KAYE, S. & PHILLIPS, C. R. 1949 The sterilizing action of gaseous ethylene oxide IV. The effect of moisture. *American Journal of Hygiene* **50**, 296–306.

LEWIS, J. C., SNELL, N. S. & ALDERTON, G. 1965 Dormancy and activation of bacterial spores. In *Spores III* eds Campbell, L. L. & Halvorson, H. O. pp. 47–54, Ann Arbor: American Society for Microbiology.

LUND, B. M., GEE, J. M., KING, N. R., HORNE, R. W. & HARNDEN, J. M. 1978 The structure of the exposporium of a pigmented clostridium. *Journal of General Microbiology* **105**, 165–174.

MacKINNON, I. H. 1974 The use of inactivators in the evaluation of disinfectants. *Journal of Hygiene* **73**, 189–195.

MARSHALL, B. J., MURRELL, W. G. & SCOTT, W. J. 1963 The effect of water activity, solutes and temperature on the viability and heat resistance of freeze-dried bacterial spores. *Journal of General Microbiology* **31**, 451–460.

MEFFERD, R. B. JR. & CAMPBELL, L. L. JR. 1951 The activation of thermophilic spores by furfural. *Journal of Bacteriology* **62**, 130–132.

MOLIN, G. & ÖSTLUND, K. 1976 Dry heat inactivation of *Bacillus subtilis* var. *niger* spores with special reference to spore density. *Canadian Journal of Microbiology* **22**, 359–363.

MOSSEL, D. A. A. 1963 The rapid evaluation of disinfectants intended for use in food processing plants. *Laboratory Practice* **12**, 898–899.

MOSSEL, D. A. A. & BEERENS, H. 1968 Studies on the inhibitory properties of sodium thioglycollate on the germination of wet spores of clostridia. *Journal of Hygiene* **66**, 269–272.

MURRELL, W. G. 1969 Chemical composition of spores and spore structures. In *The Bacterial Spore* eds Gould, G. W. & Hurst, A. pp. 215–273, London: Academic Press.

MURRELL, W. G. & WARTH, A. D. 1965 Composition and heat resistance of bacterial spores. In *Spores III* eds Campbell, L. L. & Halvorson, H. O. pp. 1–24, Ann Arbor: American Society for Microbiology.

NATIONAL CANNERS ASSOCIATION 1968 *Laboratory Manual for Food Canners and Processors* Vol. 1 *Microbiology and Processing*, Ch. 10. Westport, Connecticut: AVI Publishing.

ORTENZIO, L. F. 1966 Collaborative study of improved sporicidal test. *Journal of the Association of Official Agricultural Chemists* **49**, 721–726.

ORTENZIO, L. F., STUART, L. S. & FRIEDL, J. L. 1953 The resistance of bacterial spores to constant boiling hydrochloric acid. *Journal of the Association of Official Agricultural Chemists* **36**, 480–484.

PHILLIPS, C. R. 1952 Relative resistance of bacterial spores and vegetative bacteria to disinfectants. *Bacteriological Reviews* **16**, 135–138.

QUISNO, R., GIBBY, I. W. & FOTER, M. J. 1946 A neutralizing medium for evaluating the germicidal potency of the quaternary ammonium salts. *American Journal of Pharmacy* **118**, 320–323.

REDDISH, G. F. 1950 Cited by Schmidt, C. F. 1955 The resistance of bacterial spores with reference to spore germination and its inhibition. *Annual Review of Microbiology* **9**, 387–400.

ROBERTS, T. A. 1970 Recovering spores damaged by heat, ionizing radiations or ethylene oxide. *Journal of Applied Bacteriology* **33**, 74–94.

ROBERTS, T. A., GILBERT, R. J. & INGRAM, M. 1966 The effect of sodium chloride on heat resistance and recovery of heated spores of *Clostridium sporogenes* (PA3679/S$_2$). *Journal of Applied Bacteriology* **29**, 549–555.

ROBERTS, T. A. & HITCHINS, A. D. 1969 Resistance of spores. In *The Bacterial Spore* eds Gould, G. W. & Hurst, A., pp. 611–670, London: Academic Press.

RUBBO, S. D., GARDNER, J. F. & WEBB, R. L. 1967 Biocidal activities of glutaraldehyde and related compounds. *Journal of Applied Bacteriology* **30**, 78–87.

RUSSELL, A. D. 1971 The destruction of bacterial spores. In *Inhibition and Destruction of the Microbial Cell* ed. Hugo, W. B., pp. 451–612, London: Academic Press.

SACKS, L. E. & ALDERTON, G. 1961 Behaviour of bacterial spores in aqueous polymer two-phase systems. *Journal of Bacteriology* **82**, 331–341.

ST JULIAN, G., PRIDHAM, T. G. & HALL, H. H. 1967 Preparation and characterization of intact and free spores of *Bacillus popilliae* Dutky. *Canadian Journal of Microbiology* **13**, 279–285.

SCHMIDT, C. F. 1955 The resistance of bacterial spores with reference to spore germination and its inhibition. *Annual Review of Microbiology* **9**, 387–400.

SETLOW, P. & KORNBERG, A. 1969 Biochemical studies of bacterial sporulation and germination XVII. Sulphydryl and disulfide levels in dormancy and germination. *Journal of Bacteriology* **100**, 1155–1160.

SLEPECKY, R. & FOSTER, J. W. 1959 Alterations in metal content of spores of *Bacillus megaterium* and the effect on some spore properties. *Journal of Bacteriology* **78**, 117–123.

SPAULDING, E. H., CUNDY, K. R. & TURNER, F. J. 1977 Chemical disinfection of medical and surgical materials. In *Disinfection, Sterilization and Preservation* ed. Block, S. S. pp. 654–684, Philadelphia: Lea & Febiger.

STARK, R. L., FERGUSON, P., GARZA, P. & MINER, N. A. 1975 An evaluation of the Association of Official Analytical Chemists sporicidal test method. *Developments in Industrial Microbiology* **16**, 31–36.

STUART, L. S. 1969 Testing sterilizers, disinfectants, sanitizers and bacteriostats. *Soap and Chemical Specialties* **45**, 79–80, 84–85.

SYKES, G. 1965 *Disinfection and Sterilization*, 2nd edn, London: Spon.

SYKES, G. 1970 The sporicidal properties of chemical disinfectants. *Journal of Applied Bacteriology* **33**, 147–156.

TAMIR, H. & GILVARG, C. 1966 Density gradient centrifugation for the separation of sporulating forms of bacteria. *Journal of Biological Chemistry* **241**, 1085–1090.

TOLEDO, R. T. 1975 Chemical sterilants for aseptic packaging. *Food Technology* **29**, 102, 104, 105, 108, 110–112.

TOLEDO, R. T., ESCHER, F. E. & AYRES, J. C. 1973 Sporicidal properties of hydrogen peroxide against food spoilage organisms. *Applied Microbiology* **26**, 592–597.

TRUEMAN, J. R. 1971 The halogens. In *Inhibition and Destruction of the Microbial Cell*, ed. Hugo, W. B., pp. 137–183, London: Academic Press.

WAITES, W. M. & WYATT, L. R. 1971 Germination of spores of *Clostridium bifermentans* by certain amino acids, lactate and pyruvate in the presence of sodium or potassium ions. *Journal of General Microbiology* **67**, 215–222.

WAITES, W. M. & WYATT, L. R. 1974 The outgrowth of spores of *Clostrididium bifermentans*. *Journal of General Microbiology* **84**, 235–244.

WAITES, W. M., BAYLISS, C. E., KING, N. R. & DAVIES, A. M. C. 1979 The effect of transition metal ions on the resistance of bacterial spores to hydrogen peroxide and to heat. *Journal of General Microbiology* **112**, 225–233.

WALLEN, S. E. 1976 Sporicidal Action of Hydrogen Peroxide. *Ph.D. Thesis*. University of Nebraska, Lincoln, Nebraska, USA.

WARTH, A. D., OHYE, D. F. & MURRELL, W. G. 1963 The composition and structure of bacterial spores. *Journal of Cell Biology* **16**, 579–592.

WYATT, L. R. & WAITES, W. M. 1973 The effect of hypochlorite on the germination of spores of *Clostridium bifermentans*. *Journal of General Microbiology* **78**, 383–385.

WYATT, L. R. & WAITES, W. M. 1975 The effect of chlorine on spores of *Clostridium bifermentans*, *Bacillus subtilis* and *Bacillus cereus*. *Journal of General Microbiology* **89**, 337–344.

ZUK, I. T. & PARISI, A. N. 1974 Factors affecting biological indicator shelf life. Cited by Forsyth, M. P. 1975 *Developments in Industrial Microbiology* **16**, 37–47.

Determination of Bacterial Spore Inactivation
at High Temperatures

A. G. Perkin and F. L. Davies

National Institute for Research in Dairying, Shinfield, Reading, UK

P. Neaves and B. Jarvis

Leatherhead Food R.A., Leatherhead, Surrey, UK

Celia A. Ayres and K. L. Brown

Campden Food Preservation R.A., Chipping Campden, Gloucestershire, UK

W. C. Falloon, H. Dallyn and P. G. Bean

Metal Box Ltd., Denchworth Road, Wantage, UK

The stability and safety of many food and pharmaceutical products preserved by heat processing are dependent upon adequate destruction of microbial spores. The effect of thermal sterilization depends on the temperature and time relationships reached in the product during the process used, and the kinetics of the beneficial and adverse reactions that occur. The conditions necessary for adequate heat sterilization are usually determined empirically or are theoretically based on some simplifying assumptions regarding the shape of the heating curve, such as the development of design procedures for the sterilization of canned foods by Ball (1923) and Levine (1956). In the canning industry low acid food products are processed to ensure destruction of *Clostridium botulinum* spores by heating so that the centre of the product receives a process of at least 3 min at a temperature of 121·1°C ($F_0=3$) or an equivalent sterilizing process at another temperature. In practice, to ensure commercial sterility, processes of greater severity than $F_0=3$ are generally given; this will ensure a reduction in the probable survival of *Cl. botulinum* spores by a factor of at least 10^{12} (a $12D$ process).

In recent years the use of new processes such as ultra-high-temperature (UHT) processing and aseptic filling of liquid products, notably dairy products, has become more widespread, and is directed towards production of foods with lesser nutritional changes than in traditional commercial products and towards better product economics. The extrapolation of data obtained at traditional processing temperatures (110–125°C) to higher temperatures (135–150°C) has been shown to be unreliable (Busta 1967; Miller & Kandler 1967; Burton 1970; Burton *et al.* 1977). It is therefore essential to provide industry with data on spore heat resistance at UHT process conditions in order to ensure continuing safety and stability of commercially sterilized foods.

The object of this paper is to illustrate the range of methods currently in use for thermal inactivation studies with bacterial spores, especially at high temperatures. Techniques discussed include a capillary tube method and a small scale UHT plant useful for laboratory studies, capillary bulbs that can be placed inside sealed cans of product and processed in pilot or commercial plants, alginate beads for use with liquid products in commercial plants, and reconstituted food particles for use where particulate food products are to be processed.

Capillary Tube Technique

Thin-walled capillary tubes can be used to study the heat resistance of micro-organisms. A method was used by Davies *et al.* (1977) at the National Institute for Research in Dairying to investigate the heat resistance of *Bacillus stearothermophilus* spores at ultra-high temperatures (up to 150°C), and subsequently adapted at the Leatherhead Food R.A. for use with *Cl. botulinum* spores (Neaves & Jarvis 1977; 1978).

The technique has a number of advantages. It is convenient and easy to use in the laboratory, only simple equipment is necessary, small numbers of spores are required and holding times as low as 2 s may be used.

Heat resistance data are obtained by measuring the number of viable organisms in a suspension before and after heating. From this, the reduction in viable count is calculated. At ultra-high temperatures (135–150°C) the holding time required to give a suitable reduction in count may be as short as a few seconds. The heating-up time of samples thus becomes a significant proportion of the total process time, and inaccurate data may be obtained if corrections for the heating period are not made.

The capillary tube technique includes two ways of improving the accuracy of the data obtained: (1) The use of capillary tubes with very thin walls (0·15 mm) allows rapid heat penetration into the spore suspension. (2) A computer program calculates the heating profile for any given process and relates this to the total process time. Thus, the sporicidal effect of the heating period is determined so that the experimental data, after computer correction, show the true heat resistance of the spore suspension.

Procedure

Spores are prepared by one of the standard microbiological techniques (e.g. Davies *et al.* 1977), suspended in an appropriate medium and sealed in glass capillary tubes 100 mm long × 1 mm internal diam. × 0·15 mm wall thickness (C. E. Payne and Co. Ltd., London).

The inoculated tubes are immersed in an oil or detergent bath kept at the process temperature for times of 2–150 s. At the end of the heating period the tubes are transferred rapidly to iced water. Tubes may be held using a simple holding device such as a cork or metal carrier, and manually immersed in the heating and cooling baths. After some practice accurate timings can be maintained by use of a metronome.

An apparatus designed to transfer the tubes automatically has been described by Stern & Proctor (1954). This uses a system of weights and pulleys to transfer the capillary tubes from the heating to the cooling bath. A modification to this apparatus using pneumatic jacks to effect the transfer has also been described by Cerf *et al.* (1970), and at the Campden Food Preservation R.A. the tube carrier is transferred by a spring-operated arm. The apparatus has the advantage that the time lag in operation is reproducible.

In both methods tubes may be immersed singly or in multiples by using appropriate holders. When multiple tubes are inserted into the heating bath a momentary fall in temperature occurs, so that for holding times of less than 60 s single tubes should be used.

After cooling the tubes are washed, their exterior surfaces sterilized, and the spores extracted by mechanical crushing in diluent. Viable counts are made using standard bacteriological methods. Control counts are determined using appropriately heat-activated spores, e.g. *B. stearothermophilus* spores may be heated at 100°C for 30 min and *Cl. botulinum* spores at 70°C for 10 min.

Analysis of data

The viable counts for numbers of spores before processing and for survivors are calculated. From these, *D* values are derived using the equation:

$$D = \frac{t}{\log_{10}a - \log_{10}b}$$

where t is the holding time in minutes, a is the number of organisms heated, and b is the number of survivors.

The values of D obtained by this method do not allow for the slight lag in heating up the tube and spore suspension. Use of a computer program overcomes this problem (Perkin *et al.* 1977).

Correction of the heating period to allow for the penetration of heat into the capillary and spore suspension requires a double integration of the sporicidal effect within the capillary, with variation of temperature over the volume of the suspension and over the time of heating. Before the computer can perform this integration it is necessary to know the temperature distribution within the suspension in space and time, and also the way in which the rate of thermal death varies with temperature.

Experimental data for this are obtained using capillary tubes into which micro-thermocouples are sealed. Figure 1 shows a typical heating profile when the thermocouple is placed on the centre line of the capillary tube. Using this information, an 'adjusted' holding time is derived which represents the value of the holding time which would have an equivalent effect if the heating were instantaneous. The program then calculates D values using these adjusted times (for t in the above equation) and determines the z value over a given temperature range.

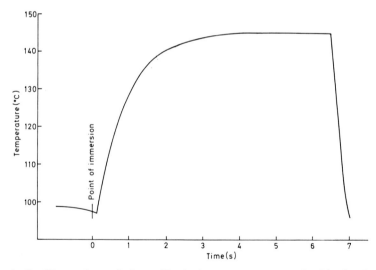

Fig. 1. Capillary tube technique. Typical temperature record with the thermocouple on the centre line of the capillary tube.

An example of the effect of computer analysis on the proportion of surviving spores for *B. stearothermophilus* is shown in Fig. 2. The use of the computer program removes the 'shoulder' from the graph but

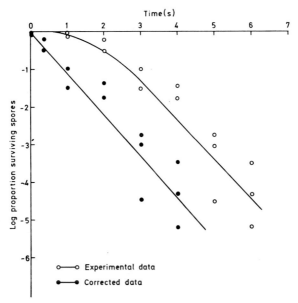

Fig. 2. Capillary tube technique. Effect of computer analysis on the thermal death curve for *B. stearothermophilus* spores at 145°C.

leaves the gradient of the linear portion unchanged. Hence a greater range of times is usable in calculating thermal death kinetics and the shoulder is due to non-instantaneous heating. Figure 3 shows the effect of computer analysis on decimal reduction times (D values) for *Cl. botulinum* spores. Decimal reduction time data are curvilinear over the temperature range 85–160°C. In other words, the z value increases as temperature increases. Following computer analysis the degree of curvature is reduced but the curve is not entirely straightened. Thus, the increase in z value with increasing temperature is partially but not entirely an artefact of the heating lag.

Over a wide range of experimental temperatures the Arrhenius relationship would be expected to be more satisfactory in describing the variation of thermal death with temperature. Figure 4 shows the effect of computer analysis on decimal reduction time for *B. stearothermophilus* spores assuming an Arrhenius relationship. Again an increase in z value with increase in temperature is indicated.

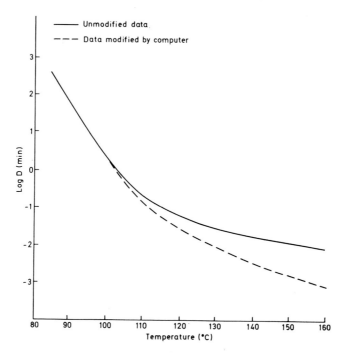

Fig. 3. Capillary tube technique. Thermal death curves for spores of *Cl. botulinum* type A (NCTC 7272).

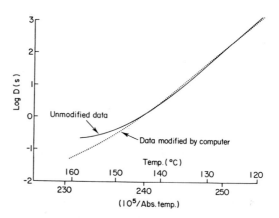

Fig. 4. Capillary tube technique. Thermal death curves for spores of *B. stearo-thermophilus*.

Laboratory-scale Product-into-steam UHT Plant

A laboratory-scale plant has been designed to process small volumes (300–500 ml) of product at UHT conditions (130–150°C for a few seconds; Perkin 1974). Results obtained using milk inoculated with spores of *B. stearothermophilus* processed at temperatures from 133–148°C for 3 s show that the plant gives the product a similar heat treatment to larger commercial plants (Perkin 1974). The use of this laboratory-scale plant has several advantages over pilot or commercial-scale plants. Speed and versatility of operation allow many processing variables to be applied to single product batches. The use of small volumes of product make it economical to use higher initial spore counts in the raw product, and a more accurate assessment of severe processing conditions is possible since correspondingly higher numbers of spores survive for counting. The plant can be isolated from processing areas thus precluding contamination, and simple modifications such as the inclusion of filters in the outlet vacuum line enable its use with dangerous pathogens such as *Cl. botulinum*.

Procedure

A schematic diagram of the sterilizer is shown in Fig. 5. Raw product is

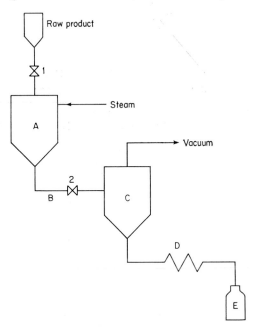

Fig. 5. Laboratory-Scale product-into-steam UHT plant. Schematic diagram.

injected into steam in vessel A where it is heated very rapidly to a temperature of 130–150°C dependent on the steam pressure. After holding in tube B for 2–5 s the product is cooled by evaporative cooling in vessel C and by heat exchange in cooling coil D before being aseptically collected in container E. A timer is used to control valves 1 and 2 to ensure reproducibility of the short holding times. A detailed description of the plant and its operation is given by Perkin (1974). Typical results for spores of *B. stearothermophilus* suspended in water are given in Fig. 6.

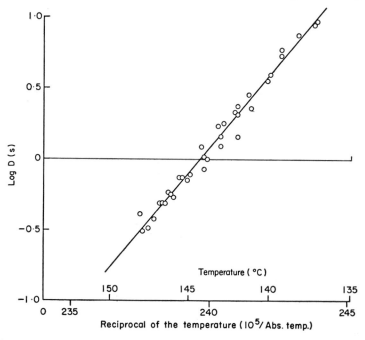

Fig. 6. Laboratory-scale product-into-steam UHT plant. Typical results for spores of *B. stearothermophilus* suspended in water.

Capillary Bulb Technique

The process given by a continuous cooker can be estimated using a spore suspension of *B. stearothermophilus* of known concentration and heat resistance, sealed in small glass capillary bulbs or tubes located inside cans of product. After processing the survivor count from the bulbs is compared with a survivor curve at 121·1°C. This technique has been used successfully in factory trials for checking processes in hydrostatic cookers (Thorpe & Grey 1971). A similar method is described by Michiels (1972) using a most probable number approach.

Spores of *B. stearothermophilus* are generally used as the biological indicator because they fulfil many of the criteria for the ideal indicator. This should have high heat resistance and relative ease of sporulation and recovery. Ideally decimal reduction times at 121·1°C should be such that a reasonable number of survivors are detectable after a particular process in order that a process estimate within fairly narrow limits may be obtained. The spore concentration and resistance should be chosen so that the process to be assessed lies on the straight line portion of a survivor plot. Usually processes in continuous cookers have been calculated using simulators so that this can be done beforehand. Spores with a thermal death time coefficient or z value of between 9 and 11°C (as determined in the particular heating and recovery system) should be chosen since processes are generally calculated using a z value of 10°C for destruction of *Cl. botulinum* spores. It is essential, therefore, that care is taken in the selection of a spore stock for a particular process if the best results are to be obtained.

Procedure

Spores of *B. stearothermophilus* suspended in buffer are sealed in glass capillary tubes or bulbs. Accurate reproducible filling is essential to reduce variability.

Survivor curves are obtained by heating the filled tubes or bulbs in an oil bath at a minimum of three temperatures, one of which is 121·1°C. The D value at each temperature is determined and the z value calculated.

If the D and z values are of the right order, the particular spore suspension is processed in cans of product in a static retort to check that the process estimation agrees with the process value obtained when using thermocouples. Tubes of spore suspension are placed along the central axis of the can while the bulbs are located on polypropylene sticks, being held in place by a piece of silicone rubber (Fig. 7). Ten cans are usually prepared in this way. Thermocouples are positioned in the same cans next to the spore suspension. The cans are seamed and processed to give a final F_0 value around that expected from the hydrostatic cooker. After cooling the cans are opened and the tubes or bulbs of spore suspension removed and washed. Spores are recovered by crushing into diluent and plated on appropriate media to enumerate survivors.

The process F_0 is read directly from the previously obtained oil bath survivor curve at 121·1°C. The results are then compared with the thermocouple data, and if good agreement is obtained it can be assumed that the estimation of the process in the hydrostatic cooker will be reasonably accurate.

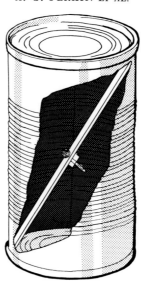

Fig. 7. Capillary bulb technique. Location of spore bulbs within a can using a polypropylene support.

To estimate the hydrostatic cooker process, batches of 10 cans containing product and the tubes or bulbs of spore suspension are sent through the cooker. Temperatures in the various legs of the cooker are noted, together with the duration and time of any stoppages. By using several batches of cans it is usually possible to obtain at least one batch which has not experienced stoppages in the steam chamber. Survivors are recovered as in the static process, and the F_0 read from the calibration experiment at 121·1°C.

Typical results from a factory trial are presented in Table 1. There is close agreement between spores and thermocouples in the control runs in the factory static retort, and it is therefore assumed that the spore results from the hydrostatic cooker provide a close estimate of the actual process.

Alginate Bead Technique

A technique in which spores of a suitable test organism (e.g. *B. stearothermophilus* str. Th 24) can be immobilized and passed through a UHT processing plant with the product, has been developed and tested under semi-commercial conditions using a scraped surface heat exchanger (Bean *et al.* 1979).

Mathematical models have been developed to predict lethal effects of UHT processing systems for particles in fluids (de Ruyter & Brunet

TABLE 1

Typical results of a factory trial using capillary bulbs sealed in 16 z cans of butter beans

		Factory static retort			
Thermocouples				Spores	
F_o values		F_o mean	No. bulbs	F_o range	F_o mean
14·5 14·9 14·9		14·8	11	14·0–15·6	14·9

Hydrostatic cooker (Mitchell Webster processing 15 min at 122·2°C)				
Trial no.	No. bulbs	F_o range	F_o mean	S.D.
1	10	14·1–16·7	15·7	0·9
2	10	14·5–16·9	15·8	0·9
3	10	13·8–16·2	15·1	1·0

1973; Manson & Cullen 1974), which serve as "rule of thumb" methods of determining process times and temperatures for a particular product. In order to ensure process safety confirmation of these conditions is required, and the alginate bead technique enables data to be obtained under normal operating conditions in the actual UHT plant employed.

Procedure

Calcium alginate beads containing homogeneously distributed spores of the test organism are formed as described by Dallyn *et al.* (1977). Different sizes of bead can be obtained by choosing suitable needle or pipette sizes during the formation process, e.g. a 10 ml disposable pipette (Sterilin Ltd) produces beads with a diameter greater than 2 mm which are ideally suited for UHT scraped surface heat exchanger studies. Using a well mixed suspension of spores of the test organism in 3% sodium alginate (Manucol DM; as supplied by Alginate Industries Ltd), beads of calcium alginate are formed by the dropwise addition of this mixture into a sterile 2% (w/v) aqueous solution of calcium chloride ($CaCl_2$). Once formed, the beads are allowed to 'cure' in the $CaCl_2$ for at least 18 h at 22–25°C. Washed beads may be stored moist in sealed, sterile containers for at least 12 months without loss of viability (Dallyn *et al.* 1977).

Spore beads may be introduced into the UHT plug valve appropriately sited in the product stream prior to the heating section and collected from the product on a series of mesh filters having a mesh size determined by the diameter of the bead used. A schematic arrangement of the relevant section of a UHT plant is shown in Fig. 8. Spores can be

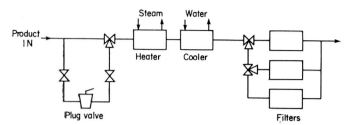

Fig. 8. Alginate bead technique. Schematic diagram of a scraped surface heat exchanger.

recovered from the beads by dissolution in 1% potassium or sodium citrate. Warming (45°C) and shaking enhance dissolution. Once released from the beads, viable surviving spores can be enumerated in a suitable medium. The pH value of the medium requires adjustment such that the final pH is optimal for the recovery of the test strain.

Figure 9 shows the distribution of numbers of surviving *B. stearothermophilus* spores in alginate beads passed through a scraped surface heat exchanger (Bean *et al.* 1979). In this experiment single heating and cooling barrels were used, and the time in the holding section was constant irrespective of the flow rate. Surprisingly, significant reductions in

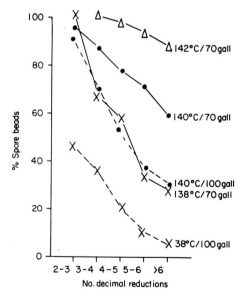

Fig. 9. Alginate bead technique. Distribution of numbers of *B. stearothermophilus* spores in alginate beads passed through a scraped surface heat exchanger (Bean *et al.* 1979).

maximum spore survival only occurred when both the flow rate was reduced and the process temperature was increased. This result indicates the need to carry out *in situ* evaluation of UHT processes.

Reconstituted Food Particle Technique

Although UHT processing and aseptic packaging of non-viscous liquids such as milk is well established, it is only comparatively recently that interest has developed in the aseptic processing of particulate food products. Spore destruction in particulates is more complicated since the transfer medium generally heats by convection, and the particles by conduction. Equipment is now available for the aseptic packaging of larger particles, e.g. peas and diced root vegetables. A method, based on the alginate bead technique already described, for the incorporation of spores into larger particles has been developed (Brown & Ayres 1978a) to assess the sporicidal efficiency of this equipment.

Particles are formed by using a purée of product to which is added a mixture of sodium alginate and sodium citrate. After adding an aqueous solution of bacterial spores to give a concentration of 10^6 spores per particle and a calcium chloride setting agent, the mixture is poured into moulds. To study the lethal effect of the heat treatment for the whole particle of food, it is important to obtain an even distribution of spores throughout the particle. Particles made by this method have been stored at 4°C for five days without changes in their properties or spore counts.

An experimental test rig used to study the effect of heat treatment on spores of *B. stearothermophilus* in potato particles (Brown & Ayres 1978b) is shown in Fig. 10.

Spores are recovered from heated particles by adding sodium citrate after first roughly macerating the particles. When plating out suspensions, an inhibitory effect on the germination of the spores may be caused by carry over of citrate. It is then necessary to add calcium to the recovery medium.

Figure 11 shows typical results for *B. stearothermophilus* spores in 1·8 cm³ potato-alginate cubes treated at 120°C compared with predicted results. The calculated results were obtained using the following equation (Stumbo 1973):

$$F_s = F_c + D_{121 \cdot 1} \cdot \log_{10} \frac{D_{121 \cdot 1} + 11 \cdot 7317 \, (F\gamma - F_c)}{D_{121 \cdot 1}}$$

Thermal death data were obtained from spores in glass capillary tubes heated in an oil bath and heat penetration data from thermocouples positioned inside particles to monitor the rate of heating. Factors which

Fig. 10. Reconstituted food particle technique. Experimental test rig.

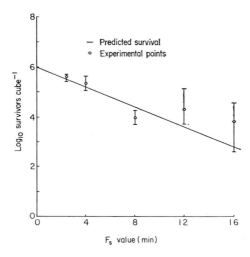

Fig. 11. Reconstituted food particle technique. Survival curve for *B. stearothermo-philus* spores in potato-alginate cubes (1·8 cm³) at 120°C.

could affect the calculated survival include shrinkage of the particle during processing, changes in thermal diffusivity, uneven spore distribution and a non-linear z value.

Symbols and Definitions

D Time required to destroy 90% of the spores or vegetative cells of a given organism. Numerically equal to the number of minutes required for the survivor curve to traverse 1 log. Mathematically equal to the reciprocal of the slope of the survivor curve.

$D_{121\cdot1}$ D value at a temperature of 121·1°C.

F Equivalent time in minutes at 121·1°C of all heat considered with respect to its capacity to destroy spores or vegetative cells of a particular organism.

F_0 F value when $z=10$°C.

F_c F_0 value of all lethal heat received by the geometrical centre of a container of food during a process.

F_s Integrated lethal value of heat received by all points in a container during a process. It is a measure of the capacity of a heat process to reduce the number of spores or vegetative cells of a given organism per container.

$F\gamma$ The equivalent in minutes at 121·1°C of all the lethal heat received by any point in a particle except the centre.

z Degrees Celsius (or equivalent in Fahrenheit) required for the thermal death curve to traverse 1 log. Mathematically equal to the reciprocal of the slope of the thermal death curve.

The authors acknowledge financial support from the Ministry of Agriculture, Fisheries and Food for the aliginate-food particle and *Clostridium botulinum* studies; and for the invaluable assistance of their colleagues.

References

BALL, C. O. 1923 Thermal process time for canned food. *Bulletin of the National Research Council, US* **7**, 1–6.

BEAN, P. G., DALLYN, H. & RANJITH, H. M. P. 1979 The use of alginate spore beads in the investigation of UHT processing. In *Proceedings of the International Meeting on Food Microbiology and Technology*, April 1978. Eds Jarvis, B. *et al.* pp. 281–294, Parma: Medicina Viva.

BROWN, K. L. & AYRES, C. A. 1978*a* Microbiological aspects of aseptic packaging. *Campden Food Preservation Research Association Technical Memorandum* No. 186.

BROWN, K. L. & AYRES, C. A. 1978*b* Microbiological aspects of aseptic packaging. *Campden Food Preservation Research Association Technical Memorandum* No. 203.

BURTON, H. 1970 Comparison of milks processed by the direct and indirect methods of UHT sterilization, III. A note on the results of spore destruction obtained with an experimental UHT milk sterilizer. *Journal of Dairy Research* **37**, 227–231.

BURTON, H., PERKIN, A. G., DAVIES, F. L. & UNDERWOOD, H. M. 1977 Thermal death kinetics of *Bacillus stearothermophilus* spores at ultra high temperatures III. Relationship between data from capillary tube experiments and from UHT sterilizers. *Journal of Food Technology* **12**, 149–161.

BUSTA, F. F. 1967 Thermal inactivation characteristics of bacterial spores at ultra high temperatures. *Applied Microbiology* **15**, 640–645.

CERF, O., GROSCLAUDE, G. & VERMEIRE, D. 1970 Apparatus for the determination of heat resistance of spores. *Applied Microbiology* **19**, 696–697.

DALLYN, H., FALLOON, W. C. & BEAN, P. G. 1977 Method for the immobilization of bacterial spores in alginate gel. *Laboratory Practice* **26**, 773–775.

DAVIES, F. L., UNDERWOOD, H. M., PERKIN, A. G. & BURTON, H. 1977 Thermal death kinetics of *Bacillus stearothermophilus* spores at ultra high temperatures I. Laboratory determination of temperature coefficients. *Journal of Food Technology* **12**, 115–129.

DE RUYTER, P. W. & BRUNET, R. 1973 Estimation of process conditions for continuous sterilization of foods containing particulates. *Food Technology* **27**, 44–51.

LEVINE, S. 1956 Determination of the thermal death rate of bacteria. *Food Research* **21**, 295–301.

MANSON, J. E. & CULLEN, J. F. 1974 Thermal process simulation for aseptic processing of foods containing discrete particulate matter. *Journal of Food Science* **39**, 1084–1089.

MICHIELS, L. 1972 Methode biologique de determination de la voleur sterilisatrice des conserves appertisées. *Industries Alimentaires et Agricoles* **89**, 1349–1356.

MILLER, I. & KANDLER, O. 1967 Temperatur und Zeit-Abhangigkeit der Sporenabtotung im Bereich der Ultrahocherhitzung. *Milchwissenschaft* **22**, 686–691.

NEAVES, P. & JARVIS, B. 1977 Thermal inactivation kinetics of bacterial spores at ultra high temperatures with particular reference to *Clostridium botulinum*. *Leatherhead Food R.A. Research Report* No. 280.

NEAVES, P. & JARVIS, B. 1978 Thermal inactivation kinetics of bacterial spores at ultra high temperatures with particular reference to *Clostridium botulinum*— second report. *Leatherhead Food R.A. Research Report* No. 286.

PERKIN, A. G. 1974 A laboratory scale ultra high temperature milk sterilizer for batch operation. *Journal of Dairy Research* **41**, 55–63.

PERKIN, A. G., BURTON, H., UNDERWOOD, H. M. & DAVIES, F. L. 1977 Thermal death kinetics of *Bacillus stearothermophilus* spores at ultra high temperatures II. The effect of heating period on experimental results. *Journal of Food Technology* **12**, 131–148.

STERN, J. A. & PROCTOR, B. E. 1954 A micro-method and apparatus for the multiple determination of rates of destruction of bacteria and bacterial spores subjected to heat. *Food Technology* **8**, 139–143.

STUMBO, C. R. 1973 *Thermobacteriology in Food Processing* 2nd edn. New York and London: Academic Press.

THORPE, R. H. & GREY, K. A. 1971 A comparison of methods for estimating process values of canned foods with particular reference to hydrostatic sterilizers. *The Fruit and Vegetable Preservation Research Association, Chipping Campden, Glos. Technical Note* No. 138.

A Standardized Thermophilic Aerobic Spore Count Applied to Raw Materials for Canning

D. W. LOVELOCK

H. J. Heinz Co. Ltd., Hayes Park, Hayes, Middlesex, UK

The primary responsibility and aim of the canner must be to produce a safe, wholesome and organoleptically acceptable product and, in order to achieve this aim, the sterilizing process must satisfy three requirements. Firstly, it must destroy the spores of the most heat resistant pathogen, *Clostridium botulinum*; secondly, it must reduce to an acceptable level the numbers of non-pathogenic organisms which might result in spoilage of the pack under the most adverse conditions of storage and distribution; thirdly, it must retain to the maximum possible extent the desirable nutritive and organoleptic properties of the food. It is generally accepted that organisms die logarithmically during heat sterilization and, for this reason, absolute sterility can never be achieved. Hence, the concept of "commercial sterility" was developed which requires that the numbers of organisms present in the food are reduced to such a level that the probability that a sterilized can will contain a viable organism represents an acceptable commercial risk.

Since the D_{250} value of *Cl. botulinum* spores is of the order of 0·2 min a sterilizing process which achieves a F_0 value equivalent to 3 min at 121·1°C will result in a probability of survival of *Cl. botulinum* spores of less than 1 in 10^{12} cans. This F_0 value is generally accepted throughout the industry and is the so-called "botulinum cook". However, the D_{250} value of *Bacillus stearothermophilus* spores, the "flat sour" organism, is considerably greater than that of *Cl. botulinum*. Although D values are very dependent upon the nature of the substrate in which the determinations are made, as a general guide, values in the range 3–6 min are commonly found in our laboratories. Therefore, even if one assumes the lowest value in this range it is evident that a sterilizing process sufficient to achieve "commercial sterility" with respect to *B. stearothermophilus* must achieve a "botulinum cook", probably with a high degree of over-kill, at the same time. Processes designed along these lines would, therefore, achieve the first two requirements stated above but, in order to achieve the third requirement, it is normally necessary to keep processing times and temperatures to a minimum. The technologist designing sterilizing processes is, therefore, faced with the necessity to give a sufficient heat treatment to kill most of the organisms in the food without giving so much heat that the desirable organoleptic properties of the food are destroyed. The control of heat resistant organisms in raw materials used for canning is, therefore, of great importance if the canner is to achieve his aim of marketing a safe, wholesome, organoleptically acceptable product.

It is our experience that the thermophilic aerobic spore count is a method which is particularly susceptible to slight variations in culture medium, heat treatment and procedure and any numerical limit applied to such tests for the purposes of acceptance or rejection of consignments of ingredient materials must take account of this fact. It is our contention that if sterilization processes are to be designed on the understanding that the ingredient materials contain not more than a certain permitted number of thermophilic aerobic spores per gram, then a carefully standardized test procedure must be conscientiously applied if the results of the tests are to be properly meaningful. The procedure developed in our laboratories and applied to raw materials for canning for more than 10 years is described below.

Media

Yeast Dextrose Tryptone Agar (Shapton medium)—first described by Shapton & Hindes (1963) and subsequently modified to use dehydrated Lab-Lemco and yeast extract.

Formula (1^{-1}): Lab-Lemco powder (Oxoid L29), 2·4 g; bacteriological peptone (Oxoid L37), 5·0 g; dextrose (Oxoid L71), 1·0 g; tryptone (Oxoid L42), 2·5 g; yeast extract powder (Oxoid L21), 0·72 g; bromocresol purple (1% aqueous soln), 2·5 ml; agar (Oxoid No. 1), 15·0 g.

Preparation: Dissolve the Lab-Lemco, peptone, dextrose, tryptone and yeast extract in the distilled water by gentle heating and stirring. Allow to cool to room temperature and adjust the pH of the solution to 7·4 by the addition of 2·5 mol 1^{-1} sodium hydroxide using a glass electrode pH meter. Add the agar and boil gently, preferably in a steam heated kettle until dissolved. Add the bromocresol purple and mix well. Distribute in 90 ml amounts in 5 oz Winchester bottles and sterilize in an autoclave at 121·1°C for 15 min.

Modified quarter strength Ringers solution. Formula (1^{-1}): Ringers solution tablets (Oxoid BR52), 2; bacteriological peptone (Oxoid L37), 5·0 g; sodium thiosulphate, 0·2 g.

Preparation: Dissolve the ingredients in the distilled water, distribute in 100 ml amounts in 5 oz Winchester bottles and sterilize in the autoclave at 121·1°C for 15 min.

The Standardized Procedure

Weigh 20 g of the material to be tested into a sterile, wide-mouthed sample jar, add 80 g of modified quarter strength Ringers solution and shake well to produce a uniform suspension. Within 15 min of making the suspension, pipette 10 ml into a bottle of molten yeast dextrose tryptone agar. The temperature of the molten medium must be not less than 80°C and not more than 1½ h must elapse between the time the medium is placed in the steamer to melt and the time the last bottle is used. Immediately after inoculation place the bottles into a steamer and steam for 2 h at 100°C. Remove the bottles from the steamer and allow to stand on a wooden or plastic bench top for 10 min. Place the bottles in a water bath at 50°C for 15 min. Remove each bottle in turn from the water bath, wipe the outside with a clean cloth, mix gently with a swirling motion and distribute the contents between four sterile Petri dishes. The plates are allowed to set for not more than 30 min, stacked, lids down, not more than eight high and incubated at 55±0·5°C for 48±8 h. After incubation, count the total colonies on four plates and report as the number of thermophilic aerobic spores per 2 g of material tested.

Discussion

The standardized procedure described above has been given without explanation for the convenience of the worker who wishes to follow the procedure but since, as was indicated earlier, small differences in medium and procedure can markedly affect the results it may be of value to discuss some of the reasons for the various operations and limitations imposed.

The numerical standard for thermophilic aerobic spores in canning raw materials used in our laboratories is not more than 30 per 2 g. This is based on the standard for food starches given by the National Canners Association (1968). Their standard states "For the five samples examined there shall be a maximum of not more than 150 spores and an average of not more than 125 spores per 10 g of starch". Since in their test, as in the above procedure, only 2 g of materials are tested and in our sampling examination scheme for some large deliveries more than five samples are examined it was considered more realistic to express the standard as per 2 g rather than per 10 g. It was also considered to be better to weigh in the diluent rather than to measure it volumetrically as recommended by the

NCA since this makes the results of tests on ingredients of widely differing densities more comparable.

The requirement that the temperature of the molten medium is not less than 80°C is to assist in ensuring that the heat treatment given is standardized. Curran & Evans (1944) and many other workers have demonstrated that sub-lethal heat treatments would induce germination in dormant spores but there has been much discussion as to the degree of lethal heat required. Shapton & Hindes (1963) described a heat treatment for this test which consisted of 20 min at 100°C followed by 10 min at 108·3°C in an autoclave but it was found in collaborative experiments with other laboratories that differences in autoclave construction, method of heating and mode of operation could drastically affect the reproducibility of results. Subsequent work in this laboratory has shown that a heat treatment of 2 h at 100°C is much more robust in the sense that it is less affected by differences in equipment and small errors in timing, and heat penetration measurements with thermocouples have shown that both heat treatments give a lethality equivalent to $F_0 = 1$. For these reasons the steamer heat treatment is now recommended rather than the original autoclave method.

The composition of the culture medium obviously has an effect on the reproducibility of results in this test but it may not be realized that this effect has two components. It is well known that the composition of the substrate in which heat treatments are carried out affects the heat resistance of the spores being heat treated. For example, Williams (1970) showed that the D_{250} value of B. stearothermophilus spores in a medium containing Oxoid agar No. 3 was twice that obtained in a medium containing Oxoid Ionagar No. 2 whilst Difco agar gave an intermediate value. However, since the heating substrate is also the culture medium one must take into account the effect of the various components on recovery and growth. It is not claimed that the medium described above gives maximum recovery of thermophilic aerobic spores from food materials. For example, our experiments have shown that the addition of soluble starch or activated charcoal will increase the count and that a slight lowering of the pH to 7·2 will markedly reduce it, but maximum recovery is not of prime importance. It is far more important in this work to obtain reproducible results so that tests done last year are comparable with similar tests done this year.

Allowing the bottles to stand on a wooden or plastic bench top for 10 min after heat treatment was originally incorporated to avoid thermal shock as the bottles are placed in the water bath. This is not such a problem with steamer heat treatment as with autoclave heat treatment but it has been known to happen. The restriction on time in the water bath before pouring is to avoid any germination and multiplication in the medium while still liquid which would obviously increase the count. A maximum of 30 min setting time was decided upon after experiments had shown that longer setting times especially in a cold laboratory resulted in reduced counts. Setting times of 1–1½ h can easily result if, for example, the plates are poured just before lunch and not incubated until after lunch. We have also experienced setting times in excess of 5 h when pouring plates was a 'last' job for the day staff and incubation of those plates a 'first' job for the night staff. It has also been our experience that counts are decreased if plates are not shielded from direct sunlight whilst setting. Stacking plates not more than eight high in the incubator ensures that all plates in a stack reach incubation temperature soon after they are put into the incubator.

After the stringent restrictions on times and temperatures in this method it may seem incongruous that such a large tolerance is allowed in incubation time. However, it has been shown that such variations in incubation time rarely affect the numbers counted, merely the size of the colonies and thus the ease with which they are counted. The tolerance of ±8 h was selected as much for operational reasons as for scientific reasons but it must be said that whilst an incubation time of less than about 40 h is probably undesirable an increase above 56 h to about 70 h over a weekend may normally be acceptable. Colony counting is assisted in this medium by the appearance of yellow acid halos around most colonies, particularly sub-surface colonies, which stand out well against the purple background. However, these halos have a tendency to fade as the agar cools and it is best to remove only a few plates from the incubator for counting at a time. Methods of counting are very much a matter for the individual worker in this as in many other methods but, in my opinion, the best method is to view the plates from the underside in reflected daylight against a white background.

Whilst variations in the standard method cannot be permitted without loss of reproducibility it is sometimes necessary to vary the method of sample preparation. If a food material is such that its incorporation into the medium at the concentration of 2 g in 90 ml so obscures the colonies that counting is difficult (e.g. full cream milk powder) this can be overcome by distributing the 10 ml inoculum between three, four or more separate bottles of agar. The total count on all plates which contain part of this 10 ml inoculum is then the thermophilic spore count per 2 g. It may also be the case that the food material does not give a uniform suspension by shaking in the diluent. In these cases it is desirable to macerate the sample in a blender or stomacher. Dried seed materials (e.g. peas, beans) or dry pasta should not be macerated, merely shaken well in the diluent since the majority of thermophiles are on the surfaces of these materials. Some materials (e.g. certain powdered gums and dehydrated ingredients) are not pipettable as a 20% suspension. Since the standard applied to this method requires that 2 g of material be tested this problem may be overcome by making a weaker initial suspension and inoculating correspondingly more bottles of agar. For example, if the strongest manageable suspension was found to be 5%, 4×10 ml inocula would be used.

The test described in this paper is used and has been successfully used for many years in this Company for the examination of a large class of raw materials for canning and, in conjunction with a properly designed sampling inspection scheme, has proved invaluable in allowing safe sterilizing processes to be designed which preserve to the maximum extent the desirable nutritive and organoleptically acceptable properties of its products.

References

CURRAN, H. R. & EVANS, R. F. 1944 Heat activation inducing germination in the spores of thermophilic aerobic bacteria. *Journal of Bacteriology* **47,** 437.

NATIONAL CANNERS ASSOCIATION 1968 *Laboratory Manual for Food Processors and Canners* 3rd edn. Westport, Connecticut: The AVI Publishing Company, Inc.

SHAPTON, D. A. & HINDES, W. R. 1963 The standardization of a spore count technique. *Chemistry and Industry* No. 6, 230–234.

WILLIAMS, M. L. B. 1970 The effect of the agar in solid media for enumerating heated spores. *Canadian Institute of Food Technology Journal* **3,** 118–119.

Inactivation of Microbial Contaminants in Irradiated Powdered or Pelleted Products

D. J. W. Barber and A. Tallentire

Department of Pharmacy, University of Manchester, Manchester, UK

F. J. Ley

Irradiated Products Limited, Swindon, UK

Ionizing radiation in the form of γ-rays from the radioisotope cobalt 60 is used in the medical products industry for sterilization purposes. Application of the process is being extended beyond 'single-use' devices such as plastic syringes, to powdered or pelleted solid materials where the microbiological background is quite different. The solid materials under investigation originate from different sources including mineral, e.g. talcum powder or kaolin, vegetable, e.g. maize starch or lycopodium powder, animal, e.g. pancreatin, bone and fish meal, or synthetic, e.g. certain pharmaceutical preparations.

The resistance of micro-organisms to radiation varies with different species and their response is influenced by the environmental conditions prevailing during irradiation. These conditions will be different for different materials, varying, for example, in terms of chemical composition, water content and bulk density. Experiments have been performed therefore on the naturally contaminated materials themselves to obtain information on the numbers and nature of the microflora originally present, as well as the overall response of innate microbial contaminants to increasing doses of γ-radiation.

Materials and Methods

Powdered and pelleted solids

Sterilizable maize starch B.P.

This grade of starch, obtained from Arbrook Ltd., Livingston, West Lothian, is Indian corn starch treated with epichlorhydrin to render it suitable for sterilization by autoclaving or prolonged steam treatment.

Lycopodium

This powder was obtained from L.R.C. Products Ltd., London. It consisted of microspores of the club moss *Lycopodium* spp., occurring as a pale yellow highly mobile powder with a fat content of around 50%.

Talcum

Talcum powder was supplied by Frederick Allen and Sons (Chemicals) Ltd., London and it consisted principally of native magnesium silicate. According to the suppliers it met the standards laid down in the 1977 Addendum to the British Pharmacopoeia, 1973 under the monograph for Purified Talcum; it had not been subjected to a sterilization treatment.

Laboratory animal diet

Pelleted diet was 'PRD' grade of Labsure Animal Foods, Poole, Dorset. It consisted of a mixture of materials derived from animal, vegetable and synthetic origins. A list of the ingredients and an outline of the operations involved in production appear in Halls & Tallentire (1978).

Recovery and enumeration of microbial contaminants

For enumeration of viable micro-organisms present on irradiated or un-irradiated solids, a necessary first step was the acquisition of cells in aqueous suspension. Such suspensions were obtained as follows: (a) powders—1 g was mixed with 10 ml sterile 0·1% peptone water by agitation on a 'Vortex-Genie' (Scientific Industries, Springfield, Mass., USA) for 30s, except for lycopodium which, because of the high surface fat content of the particles, was mixed with peptone water containing 0·1% Tween-80; (b) diet pellets—1 20 g sub-lot was blended for 30s with 200 ml 0·1% peptone water in a polyethylene bag with the aid of a Colworth 'Stomacher 400' (A. J. Seward, London). Generally, appropriate dilutions were made in sterile 0·1% peptone water so that when either 0·2 ml or 1 ml quantities of suspension were plated out and incubated, colony counts of between 60 and 100 per plate were obtained. Plating conditions varied according to the types of organisms being enumerated.

Viable aerobic micro-organisms

Replicate quantities of suspensions were used to prepare, in quadruplicate, pour plates of Plate Count Agar (Oxoid Ltd.). Incubation was at 37°C for 18 h, 32°C for 48 h or 25°C for 4 days, whichever was appropriate.

Aerobic endospores

The suspension was heated at 75°C for 15 min prior to plating out as above, followed by incubation at 37°C for 18 h.

Moulds and yeasts

Samples were plated in triplicate on the surface of rose bengal-chlortetracycline agar (Jarvis 1973) for the selective isolation and enumeration of yeasts and moulds. Colonies were counted after incubation at 25°C for between 4 and 7 days.

Viable anaerobic bacteria

Blood agar plates (blood agar base (Oxoid), plus 7% defibrinated horse blood) dried at 37°C for 30 min, were used for counts of anaerobic micro-organisms. Appropriate volumes were surface spread and plates were then incubated at 37°C for 3 days in anaerobic culture jars (Willis 1969).

Pseudomonads

Samples were used to prepare pour plates with Pseudosel Agar (Becton, Dickinson Ltd.) for isolation of *Pseudomonas* spp. Plates were incubated at 25°C for 7 days.

Presumptive coliforms

The estimation of presumed coliform organisms was done on suspensions derived from admixture of the solids and diluent and 10-fold dilutions of them. One lot of 50 ml, five lots of 10 ml, five lots of 1 ml and five lots of 0·1 ml of each suspension were used to inoculate tubes of MacConkey Broth Purple (Oxoid Ltd.) in accordance with the multiple tube fermentation technique (Anon. 1965). Tubes showing acid and gas production on incubation at 37°C for 48 h were taken as presumptive indication of the presence of coliforms. Estimates of numbers of coliform contaminants were obtained from a table of Most Probable Numbers (MPN).

γ-Irradiation of solids

The source of γ-rays was a ^{60}Co 'HOTSPOT' Mk. IV having a dose-rate of about 3·5 krads min^{-1}, measured by means of ferrous sulphate dosimetry in a geometry identical to that used for irradiating solids. The temperature in the irradiation field was approx. 27°C.

Generally solids were irradiated in contact with air in 10 g quantities contained individually in cylindrical glass vessels (50 mm diam., 65 mm height) open to the atmosphere. Samples of talcum powder were also irradiated in contact with nitrogen. For the production of the anoxic condition, 1 g amounts of powder were used and the air above and in these was removed by evacuation to a reduced pressure of 10^{-2}mm Hg ($\simeq 1.33$ Pa) in specially-designed glass vessels attached to a manifold. Evacuation was followed by admission of oxygen-free nitrogen into the vessels to atmospheric pressure. A two-fold repeat of the evacuation/N_2 cycle ensured anoxia, and powder samples were finally sealed under N_2 in the vessels.

Samples of solids were exposed to each of a number of graded doses of γ-radiation. After irradiation estimates were made of the number of viable aerobic micro-organisms or bacterial endospores surviving, according to one or other of the methods of recovery and enumeration outlined above. A dose/ln survival curve was constructed which depicted the response to radiation of the type of organism under investigation.

Deliberate contamination of talcum powder with bacterial endospores

To permit the study of a possible influence of talcum per se on the radiation response of organisms, Bacillus megaterium spores (ATCC 8245) were purposely distributed throughout this solid. A 0·15 ml quantity of a suspension containing about 10^8 B. megaterium spores ml $^{-1}$ was mixed thoroughly with 15 g talcum. After exposure to a sterile laminar air-flow to effect partial drying, the powder was further mixed by tumbling for 24 h in a closed vessel. There resulted a homogeneous distribution of spores throughout the powder at a level of around 10^6 viable spores g $^{-1}$.

Results and Discussion

Distribution of microflorae on solids

In order to indicate the degree of homogeneity of microbial contamination throughout each solid, between six and 10 samples were withdrawn from a given powdered or pelleted solid and examined individually for numbers of viable aerobic micro-organisms (37°C incubation). Counts derived from different blended suspensions, prepared from a particular solid and given the same degree of dilution, were compared by an analysis of variance. The assumption was that the variation observed within plate counts was random and comparable to that seen between mean counts derived from suspensions of like dilutions. Table 1 shows that for a given solid the microbial contamination was distributed with a reasonable degree of homogeneity, a condition which allowed valid inter-

TABLE 1

Levels of probability at which differences in variations of plate counts within and between replicate samples are significant

Solid	P
Maize starch	>0.05
Lycopodium	$0.01-0.05$
Talcum	>0.05
Animal diet	>0.05

comparisons to be made between results generated from different samples withdrawn from that bulk solid.

The innate microflora

Table 2 lists the levels of particular categories of micro-organisms sought on the four different products. For maize starch, talcum and lycopodium powders, estimates of viable aerobes derived from plate counts resulting from incubation at 25°C do not differ markedly from corresponding levels obtained on incubation at 37°C, showing that there is no pre-dominance of micro-organisms favoured by one particular incubation condition. This contrasts with the animal diet for which the count obtained at 25°C is five-fold greater than that seen with 37°C incubation.

TABLE 2

The extent and types of micro-organisms comprising the innate microflora of four powdered or pelleted products of different origins

Type of micro-organism	Maize starch g^{-1}	Talcum g^{-1}	Lycopodium g^{-1}	Animal diet g^{-1}
Viable aerobes (37°C)	2.2×10^4	4.6×10^3	3.2×10^4	1.6×10^4
Viable aerobes (25°C)	8.8×10^3	2.4×10^3	2.7×10^4	8.1×10^4
Bacterial endospores	8.0×10^3	1.4×10^3	2.5×10^4	9.9×10^3
Moulds and yeasts	1.0×10^1	8.7×10^1	—	7.3×10^2
Presumptive coliforms	0	3.0×10^1	—	4.0×10^1
Pseudomonads	0	0	—	0
Anaerobes	—	3.3×10^1	—	—

The 'PRD' diet has a high cereal content that possesses known high levels of contamination made up principally of soil-borne micro-organisms (Halls & Tallentire 1978) and so the wide difference in counts is probably to be expected. Soil-borne organisms could also have contributed significantly to the original microflorae of maize starch, talcum and lycopodium, but the present results indicate that the treatments given to these powders during the preparative stages of processing were responsible for selective reduction in the levels of this type of micro-organism.

Particularly significant is the fact that in each instance the bacterial endospore count lies close to the viable aerobic microbial count; aerobic endospores comprise 36, 30, 78 and 62% of the total microflorae of maize starch, talcum, lycopodium and animal diet respectively. Clearly, process treatments have resulted in the selection of resistant endospores.

The counts obtained for moulds and yeasts, coliforms or pseudomonads show that these categories of organisms do not make up marked proportions of the innate microflorae of the solids. The estimate of the incidence of anaerobic bacteria on talcum was carried out in view of the Pharmacopoeial description (Anon. 1972) which mentions a likely possession of numbers of *Clostridium* spp. Our finding of around 30 anaerobes per g talcum, while generally high, is in accord with the 'rule of thumb' estimate that around 1% of the total microflora is anaerobic bacteria.

Responses of the innate microflorae to γ-radiation

Figure 1 shows dose-survival curves for the microflorae of maize starch, talcum, lycopodium and animal diet, as measured by means of counts for viable aerobes or for endospores. A striking feature of this figure is the existence of two distinct shapes of survival curves amongst the four solids examined. For maize starch and talcum, the responses to radiation of the microflorae are fitted by curves that are convex to the dose axis, whereas for lycopodium and animal diet, the responses similarly scored are reasonably fitted by straight lines. In the simplest terms, these findings imply that the microflorae of lycopodium and animal diet comprise micro-organisms responding uniformly to radiation, while those of maize starch and talcum have organisms whose responses to radiation vary markedly, the organisms of low radiation response becoming increasingly predominant as survivors of radiation as dose is increased. Clearly the data in Table 2 show that all four products have innate microflorae made up of organisms of different types, although for each product bacterial endospores comprise a significant proportion of the total organisms. In such instances spores are supposed to play a major role in defining radiation response; the fact that the 'total aerobic' response and the 'endospore' response for a given product either superimpose or are parallel supports this supposition. Of course, the simplest explanations may not suffice for the complex conditions that undoubtedly exist in the

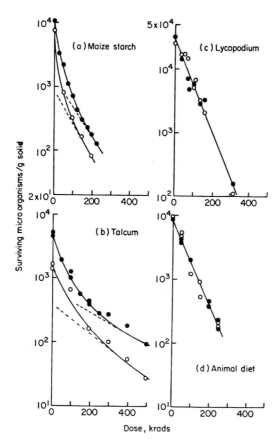

Fig. 1. Dose-survival curves from the γ-irradiation of the innate microflorae of our solid products in contact with air. Viable aerobes (●); bacterial endospores (○).

products examined here. Surprisingly, organisms on lycopodium and on animal diet, where conditions are particularly complex, exhibit the most uniform response over the survival range tested.

It is appreciated that the responses to γ-irradiation of the microflorae of the four products have been followed over a limited range of survival levels only. The limits are set by practical considerations; for the upper, the magnitude of initial contamination of the particular solid is the determinant and, for the lower, the ability to detect surviving organisms in suspensions obtained from blending the solid with a minimal volume of eluent is the governing factor. In the present work, in which plate count techniques have been used for scoring survivors of irradiation, the lower limit corresponds to a survival level of around 100 viable organisms per g of powder. This leaves ample scope for departure from linearity in the dose-survival curves of lycopodium and animal diet over levels of survival beyond this limit. To assess whether or not such departures really occur will necessitate a radical change in methodology, but even without this it is possible, in our view, to make a general analysis of the responses to radiation of the microflorae of the four solid materials under investigation.

Linear regression analysis, applied to combined data for viable aerobes and endospores for each of lycopodium and animal diet, provides an estimate of the decimal reduction dose (D_{10}) and its associated errors. Estimates are 156 ± 10 krads and 151 ± 22 krads for lycopodium and animal diet respectively. D_{10} values at and around these levels are similar to those quoted as typical for populations of bacterial endospores irradiated in air (Ley & Tallentire 1964), which clearly is in accord with the finding that the innate microflorae of these solids mainly comprise organisms of this type (Table 2). Furthermore, the essentially equivalent responses to radiation of microflorae present on two materials markedly different in chemical composition suggest an absence of any effect of the support material itself on response.

A functional measure of the radiation responses of the microflorae of maize starch and talcum, the two solid materials that exhibit dose-survival curves distinctly convex in shape, is one observed from datum points at low survival levels. Available evidence, admittedly limited, suggests that each set of terminal points falls along a linear course (*see* the broken lines in Fig. 1a & b), which, for purposes of comparison, is again characterized by a D_{10} value. For maize starch, D_{10} values were 170 and 160 krads, when measurements were done on the total viable aerobic population and aerobic endospore population respectively, and for talcum, 350 and 310 krads for corresponding conditions of measurement. The values for maize starch are similar to the estimates derived from lycopodium and animal diet (presumed to be determined by spores) whereas those for talcum signify an unusually low response to radiation. No obvious feature associated with the composition of talcum or its microflora provides an explanation for this low response, but because of the commercial application of irradiation to the inactivation of micro-organisms in talcum, further investigation was thought desirable.

Supplementary studies on γ-irradiated talcum

A simple interpretation of the convex-shaped survival curve derived from a mixed population of organisms is the increasing predominance with dose of survivors of low response to radiation. It is noted however that this curve shape can also result from a population of organisms of a given type in respect of radiation response. It occurs when conditions during irradiation are changed as a result of irradiation, from a kind which determines a high response to one that gives a low response. Most commonly in materials irradiated in air which are not undergoing continuous aeration during irradiation, the change in condition is from an oxygen-rich state to one that is oxygen-depleted, the oxygen being exhausted locally as a consequence of involvement in radiation-induced chemical processes. Although evidence for the existence of these processes in irradiated talcum is lacking, the occurrence of a convex curve, coupled with the fact that the high D_{10} values seen with this material correspond to values for endospores of aerobes irradiated in a dried condition under anoxia (Ley 1964), caused us to examine whether anoxia is playing a role in the low response associated with talcum.

Figure 2a gives the responses of aerobic endospores on talcum irradiated in contact with air and in contact with nitrogen. The overall reduction in response seen with nitrogen, occurring to an equal extent over the dose range tested, is a clear indication that induction of local anoxia is not important in fixing the radiation response of contaminants of talcum.

Another possibility is that the talcum *per se* may be responsible for the convex nature of the survival curve. Past experience with a related material, kaolin which consists mainly of aluminium silicate, would argue

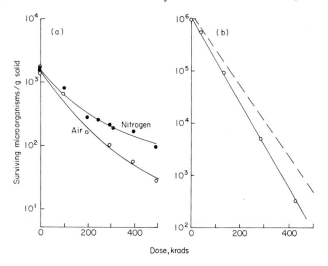

Fig. 2. Dose-survival curves from γ-irradiation of aerobic bacterial endospores present on talcum. (a) The innate endospore population in air ($\bigcirc$) and in nitrogen ($\bullet$). (b) *Bacillus megaterium* spores in talcum (—) and aerated water (- - -).

against this (Tallentire 1958), yet it was thought desirable to obtain confirmation of the inertness of talcum. The means by which this latter was done was to follow the radiation response of a reference organism, *B. megaterium* spores, distributed throughout talcum in contact with air. The strictly linear response observed over four decades of inactivation (Fig. 2b) refutes this second probability. It is interesting to compare the dose-survival curve of *B. megaterium* spores on talcum with that of spores suspended in aerated water (*see* the broken line in Fig. 2b). In the light of earlier work (Tallentire & Powers 1963), the closeness of these responses suggests that the water content of the air-dried talcum is a major environmental determinant of the level of radiation response of the contaminants on talcum.

The two pieces of evidence gathered to date provide no support for the notion that the comparatively low radiation response of the innate microflora of talcum is associated with environmental irradiation conditions peculiar to the solid support. The alternative explanation is that the microflora of talcum has a fraction of organisms that are inherently of high resistance to radiation. Our present work is aimed at recognizing the resistance parameters of these organisms, at identifying the types of organisms that possess high resistance and at determining their frequency in the innate microflora of talcum.

References

ANON. 1965 *Standard Methods for the Examination of Water and Wastewater*, 12th edn, New York: American Public Health Association Inc.

ANON. 1972 *Martindale, The Extra Pharmacopoeia*, 12th edn, p. 577, London: The Pharmaceutical Press.

HALLS, N. A. & TALLENTIRE, A. 1978 Effects of processing and gamma radiation on the microbiological contaminants of a laboratory animal diet. *Laboratory Animals* 12, 5–10.

JARVIS, B. 1973 Comparison of an improved rose-bengal chlortetracycline agar with other media for the selective isolation and enumeration of moulds and yeasts in food. *Journal of Applied Bacteriology* 36, 723–727.

LEY, F. J. 1964 The effects of ionizing radiations in producing sterility in pharmaceutical products with special reference to spores. In *Ionizing Radiation and the Sterilization of Medical Products*. pp. 51–62, London: Taylor and Francis.

LEY, F. J. & TALLENTIRE, A. 1964 Sterilization by radiation or heat—some microbiological considerations. *Pharmaceutical Journal* 192, 59–61.

TALLENTIRE, A. 1958 An observed 'oxygen effect' during gamma-irradiation of dried bacterial spores. *Nature* 182, 1024–1025.

TALLENTIRE, A. & POWERS, E. L. 1963 Modification of sensitivity to x-irradiation by water in *Bacillus megaterium*. *Radiation Research* 20, 270–287.

WILLIS, A. T. 1969 Techniques for the study of anaerobic spore-forming bacteria. In *Methods in Microbiology* eds Norris, J. R. & Ribbons, D. W. pp. 79–115, London and New York: Academic Press.

Effects of Temperature and Carbohydrate on the Growth and Survival of *Yersinia enterocolitica*

Susan A. Buckeridge, A. Seaman and M. Woodbine

Microbiological Unit, Department of Applied Biochemistry and Nutrition, University of Nottingham, School of Agriculture, Sutton Bonington, Loughborough, Leicestershire, UK

The organism now known as *Yersinia enterocolitica* has been called by a variety of names; *Pasteurella* "X" (Knapp & Thal 1963), *Pasteurella pseudotuberculosis* b (Feeley *et al.* 1976) and *Pasteurella pseudotuberculosis*-like bacterium (Frederiksen 1964). Recently, in the 8th edition of Bergey's Manual of Determinative Bacteriology (Buchanan & Gibbons 1974), it has been included in the family Enterobacteriaceae.

The organism is regularly isolated from human patients with acute abdominal pains and appendicitis-like symptoms. The portal of entry of infection is probably the gastro-intestinal tract, but in most cases the source of infection has not been defined (Sonnenwirth 1974). The presence of this bacterium in foodstuffs is potentially hazardous to human health. Isolates have already been recovered from several types of food, including pork (Zen-Yoji *et al.* 1974), ice cream (Wauters 1970), mussels (Spadaro & Infortuna 1968) and oysters (Toma 1973). Undoubtedly, isolates will be cultured from an even wider variety of foods as these are examined more extensively. The first outbreak of illness from *Y. enterocolitica* in which foodborne transmission was documented, was from chocolate milk supplied to school children in New York (Morbidity & Mortality Weekly Reports 1977).

With regard to the methods of preparation, transport and storage, and to the composition of chocolate milk, the ability of the organism to grow and survive at refrigeration temperatures and to survive moist heat stress was investigated.

Materials and Methods

Bacterium and media

Yersinia enterocolitica Strain NTCC 10598 (National Type Culture Collection, Colindale, London) was used in this study. Powder preparations for growth media were obtained from Oxoid Ltd., Basingstoke, Hampshire. Sucrose and lactose were from Fisons Scientific Apparatus Ltd., Loughborough, Leicestershire. Mineral salts and acids were obtained from BDH Ltd., Poole, Dorset.

Counting techniques

Viable plate counts were made by inoculating and evenly distributing 0·5 ml of diluted bacterial suspension over the surface of nutrient agar plates dried at 37°C for 2 h. Bacterial suspensions were diluted in quarter strength Ringers. A comparison of colony counts obtained by diluting in phosphate buffer saline or quarter strength Ringers, showed that higher counts were obtained with the Ringers. Colonies were counted after 72 h incubation at 22°C.

Growth measurements

Absorbance was measured by one of two methods: (a) using an EEL absorptiometer (Model A, Evans Electroselenium Ltd., Harlow, Essex) fitted with a 16×150 mm test tube adaptor and a neutral density filter; (b) using a Unicam SP600, Series 2, Spectrophotometer (Pye Unicam Ltd., Cambridge). Measurements were made at 650 nm on 3 ml quantities of culture with distilled water as blank and values were corrected for a sterile nutrient broth blank.

There was good correlation between total colony count and absorbance during exponential growth and for some period after maximum growth. An absorbance (650 nm) of 0·20 was equivalent to a count of approx. 3×10^8 colony forming units ml^{-1}.

Growth conditions

For tubes, 5 ml volumes of media were inoculated with 0·1 ml of an 18 h nutrient broth culture. For larger volumes, 300 ml metric medical bottles containing 100 ml of medium were inoculated with 0·5 ml of an 18 h nutrient broth culture.

Heating menstrua

All heating menstrua contained 0·02 M phosphate buffer pH 6·8, with the exception of the milk. The buffer was prepared as a 10-fold concentrated solution and autoclaved at 121°C for 15 min. Phosphate buffer saline contained 5 gl^{-1} NaCl and 0·02 M phosphate buffer. Phosphate buffer-nutrient broth consisted of nutrient broth powder (25 gl^{-1}), and phosphate buffer (0·02 M). Buffer was added after heat sterilization; UHT milk had a pH of 6·7. Sugars were added when required as dry sterilized powders (160°, 2 h).

Survival studies

Bacteria from 18 h, 22°C nutrient broth cultures in the late exponential phase of growth, were separated from the culture medium by centrifugation (12 000 **g**, 15 min, 4°C). Cells were resuspended to an absorbance value of 0·20 in the appropriate heating menstrua. Samples of cell suspension (0·5 ml) plus 4·5 ml volumes of the heating menstrua were mixed and held at the required temperature ($\pm 0·5°C$) in a water bath. The tubes were swirled periodically, 0·5 ml samples withdrawn and the number of survivors estimated as colony forming units. There was no decrease in the number of viable cells when suspensions were held at room temperature in the heating tubes for the period of the experiments.

Results

Cold temperature growth

Tube cultures of *Y. enterocolitica* were grown in nutrient broth at temperatures in the range 4–44°C. The initial cell number was approximately 10^7 colony forming units ml^{-1}. Growth was measured over a 15 day period and population doubling times (t_d) determined (Table 1). Of the temperatures used, 22°C was considered optimum, though good growth was obtained at both 10° and 4°C. The lag period, prior to the establishment of exponential growth phase, was generally 2–6 h at all temperatures, except at 4°C, when it was about 40 h. At 22° and 37°C, the addition of 1% (w/v) sucrose to the growth medium had a negligible effect on the growth of *Y. enterocolitica*. However, at 44°C, 1% (w/v) sucrose increased maximum growth to near that occurring at 37°C. In a similar manner, the growth maximum achieved at 10°C, in the presence of 1% (w/v) sucrose, was approximately equivalent to that at the optimum temperature and at 4°C, sucrose increased maximum growth considerably (Table 1).

TABLE 1

Growth of Y. enterocolitica *in nutrient broth and nutrient broth plus 1% (w/v) sucrose*

Incubation temperature (°C)	Turbidity in nutrient broth*	Turbidity in nutrient broth +1% (w/v) sucrose*
4	0·38 (>20)	0·50 (>20)
10	0·44 (14)	0·60 (12)
22	0·75 (6)	0·76 (4)
37	0·23 (6)	0·25 (4)
44	0·10 (6)	0·22 (4)

*Culture turbidity (measured as absorbance) was determined after 15 days incubation at the appropriate temperature.

Figures in parenthesis are values for population doubling times (t_d) expressed in h.

In a second experiment, 100 ml nutrient broth or nutrient broth plus sucrose (1–10% w/v), were incubated at 2°C, 4°C and 6°C (Table 2). Growth rate and maximum growth increased with increasing temperature of incubation. At 4°C and 6°C, the presence of up to 5% (w/v) sucrose had no effect on growth rate, but increased the maximum growth over that obtained in nutrient broth alone. Concentrations above 5% and up

TABLE 2

Growth of Y. enterocolitica *at low temperatures in nutrient broth with various concentrations of sucrose*

Incubation tempera- ture (°C)	Turbidity* in media containing sucrose (% w/v)					
	0	1	2	5	7	10
2	0·37 (52)	0·47 (54)	0·45 (54)	0·47 (68)	0·36 (68)	0·36 (68)
4	0·56 (36)	0·70 (36)	0·66 (36)	0·58 (48)	0·56 (48)	0·46 (60)
6	0·55 (34)	0·69 (34)	0·69 (34)	0·69 (34)	0·55 (34)	0·46 (46)

* Culture turbidity (measured as absorbance) was determined after 20 days incubation at the appropriate temperature.

Figures in parenthesis are values for population doubling times (t_d) expressed in h.

to 10% sucrose slightly inhibited growth. At 2°C, however, 1–5% (w/v) sucrose caused the growth rate to be marginally slower, but again effected an increase in maximum growth over that in nutrient broth alone. Both 10% and 7% (w/v) sucrose were inhibitory. It is noteworthy that in nutrient broth at 4°C and 22°C, growth occurred in the presence of 40% sucrose (w/v). This concentration is much higher than many bacteria can withstand and this organism might therefore be designated osmo-tolerant. However, there was no growth in nutrient broth with 65% (w/v) sucrose.

Heating studies

Heating of *Y. enterocolitica* cells at 50°C and 62°C, extended the period before colonies appeared on nutrient agar plates from 48 to 72 h. Colonies produced by the heat-treated cells varied greatly in size, the majority being smaller than corresponding non-heat-treated cells after an identical incubation period (72 h, 22°C). Colony counts did not change significantly after incubation of plates for more than 72 h. Hexagonal shaped colonies (1·00–2·00 mm diam.), were produced from the lowest dilution of cells which had been heat-treated for 30 min at 62°C in UHT milk, both with and without added sucrose. Dark zones were apparent around the hexagonal colonies. A hexagonal colony when streaked on to nutrient agar, did not give rise to hexagonal colonies. None of the colonies from heat-stressed cells had the slightly serrated edge of 'normal' non-heat-treated *Y. enterocolitica* cells grown at 22°C. Hexagonal shaped colonies have been noted by Gooder & Maxted (1961) for *Streptococcus pyogenes* grown on Hartley salt serum agar with excess surface moisture.

The effect of different heating menstrua on the survival of *Y. enterocolitica* cells is shown in Figs 1 & 2. A period, during which time little or

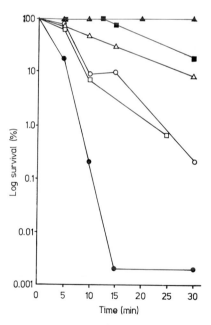

Fig. 1. Survival of Y. *enterocolitica* at 50°C in various media. ○, PBS; ●, PBS + 10% S; □, PBS + 10% L; ■, NB; △, NB + 10% S; ▲, milk + 10% S. See Table 3 for abbreviations.

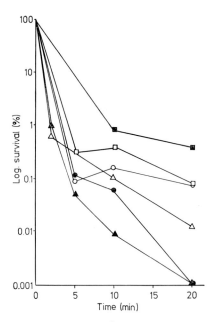

Fig. 2. Survival of Y. *enterocolitica* at 62°C in various media. ○, PBS; ●, PBS+
20 % S; □, PBS+20 % L; ■, milk; △, milk+20 % S; ▲, PBS+65 % S. See Table 3
for abbreviations.

no loss in survival occurred was evident when cells were heated at 50°C
and this varied with the heating menstruum used. Linear and semi-
logarithmic graphs of per cent survival against heating times allowed
calculation of D_{50} and $D_{62°C}$ values (Decimal reduction times). Graphs
drawn for calculation of the D values were those of the best possible
straight line through points obtained for mortalities less than 90%
(Tables 3 & 4).

TABLE 3

$D_{50°C}$ *values of* Y. enterocolitica *in various concentrations of sucrose and lactose in
phosphate buffer saline, nutrient broth and milk*

Heating menstruum*	D value in presence of sugar (% w/v)						
	0	1	2	5	10	20	40
PBS+L	9·00	—	—	—	8·25	5·00	—
PBS+S	9·00	5·75	—	4·25	1·75	—	—
NB+S	29·25	27·50	—	27·50	27·25	—	—
Milk+S	30·00	—	30·00	30·00	30·00	30·00	30·00

*PBS, phosphate buffer saline; NB, nutrient broth; S, sucrose; L, lactose.

TABLE 4

$D_{62°C}$ values of Y. enterocolitica in various concentrations of sucrose and lactose in phosphate buffer and milk

Heating menstruum*	D value in presence of sugar (% w/v)							
	0	2	5	10	20	30	40	65
PBS + L	0·65	0·80	—	0·65	1·02	0·72	—	—
PBS + S	1·57	—	—	—	1·00	—	0·95	0·92
Milk + S	5·00	—	3·75	3·75	3·25	—	3·20	—

*PBS, phosphate buffer saline; NB, nutrient broth; L, lactose; S, sucrose.

Discussion

Survival curves obtained from Y. enterocolitica cells heated at 50°C, were of two kinds. These either took the form of a shoulder (due to an initial lag in death rate), followed by a logarithmic decline which tailed, or a concave curve. The initial lag can be explained by a gradual accumulative effect of heating, resulting in injury but not loss in viability. Strange (1976) suggested that either a certain 'dose' is necessary to inactivate a cell, or that death results from several events ('multitarget model'). Clumping of cells was reported by Stumbo (1965) to explain such shoulders and this was shown to be the case for Streptococcus faecalis cells heated in milk (Hansen & Reimann 1963). Stumbo (1973) stated that salt solutions, for example phosphate buffer, often flocculate the suspended bacteria into multicellular clumps. Logarithmic death rate ensues only when the number of surviving cells per clump is reduced to one. Moats et al. (1971) have suggested tailing of survival curves may be a normal characteristic of a bacterial population. Concave curves may result from the fact that not all cells in a batch culture population are in the same stage of their division cycle (Hansen & Reimann 1963).

From $D_{50°C}$ values, it was found that nutrient broth offered a greater degree of protection to cells subjected to a heat stress, than did phosphate buffer saline. In both instances, $D_{50°C}$ values decreased as sucrose increased from 1 to 10%. Phosphate buffer saline containing 10% (w/v) lactose gave approximately five times the $D_{50°C}$ value as phosphate buffer saline containing 10% (w/v) sucrose. At 62°C, as at 50°C, increasing the sucrose concentration in phosphate buffer, led to a decrease in survival. However, lactose afforded a degree of protection to the cells and increased survival. D_{50} and $D_{62°C}$ values were considerably greater when milk was used as heating menstruum. The presence of sucrose in the milk did not appear to have any further protective effect. These observations of increased killing of *Y. enterocolitica* in the presence of sucrose, are in contrast to heat resistance studies of *Salmonella* spp. reported by several workers, for example, Goepfert *et al.* (1970) and Baird-Parker *et al.* (1970).

Corry (1974; 1976) has suggested that a solute such as a sugar or a polyol, able to pass across the cell membrane, replaces some of the cytoplasmic water. The limited plasmolysis resulting from this effect brings about a degree of protection to the cell during subsequent heating. However, a sugar unable to penetrate the cell membrane, causes more severe plasmolysis. Such a dehydration effect is more protective during heating than the limited water replacement. This, therefore, may explain why lactose, not utilized by *Y. enterocolitica* (NTCC 10598), offered a degree of protection to heated *Y. enterocolitica* cells, whereas sucrose did not. However, the degree of protection was only slight. It is unlikely, therefore, that the sugar content of a medium or food is of significant importance in increasing the survival of *Y. enterocolitica* at high temperatures. Furthermore, it is unlikely that the lactose content of milk and indeed the added sucrose in chocolate milk would afford protection to the cells on heating. The increased heat resistance of *Y. enterocolitica* in milk may have been due to a protective action exerted by various milk constituents, e.g. Ca^{2+} or Mg^{2+} (Strange 1976), salts, proteins or fats. That it may have been partly due to proteins is evident from the results obtained with nutrient broth which provided a similar protection to heat at 62°C as did milk. The negligible or even inhibitory effect of sucrose on the heat survival of *Y. enterocolitica* may be related to the ability of this organism to tolerate and grow in concentrations of sucrose up to 45% (w/v). It has been suggested that the ability of an organism to grow under low a_w is not due to the possession of a mechanism whereby the internal osmotic concentration of the cell may be regulated, but that certain enzymes or enzyme systems are themselves adapted to function under such conditions (Brown 1976). This, therefore, suggests a direct involvement of enzymes in determining the heat sensitivity and subsequent resuscitation of heat stressed cells.

Sucrose did increase the growth rate and yield of *Y. enterocolitica* at refrigeration temperatures of above 4°C. This is in accordance with the findings of Gill (1976) that the rate of bacterial growth on or in a food may depend upon the free sugar content.

The authors thank Dr H. F. Jenkinson and Mr D. Fowler for their assistance.

References

BAIRD-PARKER, A. C., BOOTHROYD, M. & JONES, E. 1970 The effect of water activity on the heat resistance of heat sensitive and heat resistant strains of Salmonellae. *Journal of Applied Bacteriology* **33**, 515–522.

BROWN, A. D. 1976 Microbial water stress. *Bacteriological Reviews* **40**, 803–846.

BUCHANAN, R. E. & GIBBONS, N. E. 1974 *Bergey's Manual of Determinative Bacteriology*, 8th edn, American Society for Microbiology, Baltimore: Williams and Wilkins Co.

CORRY, J. E. L. 1974 The effect of sugars and polyols on the heat resistance of Salmonellae. *Journal of Applied Bacteriology* **37**, 31–43.

CORRY, J. E. L. 1976 Sugar and polyol permeability of *Salmonella* and osmophilic yeast cell membranes measured by turbidimetry and its relation to heat resistance. *Journal of Applied Bacteriology* **40**, 277–284.

FEELEY, J. C., LEE, W. H. & MORRIS, G. K. 1976 *Yersinia enterocolitica*. In *Compendium of Methods for the Microbiological Examination of Foods* ed. Speck, M. L. pp. 351–357, Washington: American Public Health Association.

FREDERIKSEN, W. 1964 A study of some *Yersinia pseudotuberculosis*-like bacteria ("*Bacterium enterocoliticulum*" and "*Pasteurella X*"). In *Proceedings of the XIVth Scandinavian Congress of Pathology and Microbiology*. Oslo, 103–104.

GILL, C. O. 1976 Substrate limitation of bacterial growth at meat surfaces. *Journal of Applied Bacteriology* **41**, 401–410.

GEOPFERT, J. M., ISKANDER, I. K. & AMUNDSON, C. H. 1970 Relation of the heat resistance of Salmonellae to the water activity of the environment. *Applied Microbiology* **19**, 429–433.

GOODER, H. & MAXTED, W. R. 1961 External factors influencing the structure and activities of *Streptococcus pyogenes*. *Symposium of the Society for General Microbiology* **11**, 151–173.

HANSEN, N. H. & REIMANN, H. 1963 Factors affecting the heat resistance of nonsporing organisms. *Journal of Applied Bacteriology* **26**, 314–333.

KNAPP, W. & THAL, E. 1963 Untersuchungen über die kulturell-biochemischen, serologischen, tierexperimentellen und immunologischen Eigenschaften einer vorlaufig "*Pasteurella* X" benannten Bakterienart. *Zentralblatt fur Bakteriologie, Parasitenkunde, Infektionskrankheiten und Hygiene. Abteilungen originale* **190**, 472–484.

MOATS, W. A., DABBAH, R. & EDWARDS, V. M. 1971 Interpretation of non-logarithmic survivor curves of heated bacteria. *Journal of Food Science* **36**, 523–526.

MORBIDITY AND MORTALITY WEEKLY REPORT. February 1977. *Yersinia enterocolitica* outbreak, New York. **26**, 53–54.

SONNENWIRTH, A. C. 1974 Yersinia. In *Manual of Clinical Microbiology*. 2nd edn eds Lennette, E. H., Spaulding, E. H. & Truant, J. P. Washington: American Society for Microbiology.

SPADARO, M. & INFORTUNA, M. 1968 Bolamenta di *Yersinia enterocolitica* in *Mililus gallaprovinicialis Lamk*. *Bollettino della Società Degli Italiana di Biologia Sperimentale* **44**, 1896–1897.

STRANGE, R. E. 1976 *Microbial Response to Mild Stress*. Co. Durham: Meadowfield Press Ltd.

STUMBO, C. R. 1965 *Thermobacteriology in Food Processing*. New York: Academic Press.

STUMBO, C. R. 1973 *Thermobacteriology in Food Processing*. 2nd edn. New York: Academic Press.

TOMA, S. 1973 Survey on the incidence of *Yersinia enterocolitica* in the province of Ontario. *Canadian Journal of Public Health* **64,** 477–487.

WAUTERS, G. 1970 *Contribution à l'étude de* Yersinia enterocolitica. thèse d' agrégation. Louvain, Belgium: Vander.

ZEN-YOJI, H., SAKAI, S., MARUYAMA, T. & YANAGAWA, Y. 1974 Isolation of *Yersinia enterocolitica* and *Yersinia pseudotuberculosis* from swine, cattle and rats at an abattoir. *Japanese Journal of Microbiology* **18,** 103–105.

Survival and Growth of *Yersinia enterocolitica* in Broth Media and in Food

MARGARET KENDALL AND R. J. GILBERT

Food Hygiene Laboratory, Central Public Health Laboratory, London, UK

Sporadic cases of human gastrointestinal infection due to *Yersinia enterocolitica* have been recognized and reported with increasing frequency since 1966 (Highsmith *et al.* 1977). Workers in various countries throughout the world have recorded the occurrence of this organism (Rabson & Koornhof 1972; Zen-Yoji & Maruyama 1972; Hinderaker *et al.* 1973; Rakovský *et al.* 1973; Vandepitte *et al.* 1973; Kohl *et al.* 1976; Szita & Svidró 1976). Useful reviews of a general nature have also been published (Mair 1975; Morris & Feeley 1976; Highsmith *et al.* 1977; Bottone 1977).

At least five outbreaks in which a common food or water source was suspected are known to have occurred in the period 1973–1978. Three of these were explosive community outbreaks in Japan involving junior and primary schools. Zen-Yoji and co-workers described an incident in which 198 of 1086 pupils suffered typical symptoms of abdominal pain (76%), fever (61%), headache (60%), diarrhoea (36%), malaise (33% and vomiting (12%). Half of the patients experienced severe pain in the umbilical region or lower right quadrant and three cases underwent appendectomy. One hundred and thirty two isolates of *Y. enterocolitica* were serotype 0:3, two were serotype 0:5 and six serotype 0:9: one patient carried both types 0:3 and 0:9 (Zen-Yoji *et al.* 1973). The other two incidents in Japan (Asakawa *et al.* 1973) were similar to the one described above and affected a total of 733 children and teachers. All the strains isolated were *Y. enterocolitica* type 0:3. The fourth outbreak occurred in Czechoslovakia (Olšovský *et al.* 1975) and centred on two establishments for the collective care of children, both of which received their food from a single kitchen. In none of the outbreaks described above was the vehicle of infection identified.

The fifth outbreak (Black *et al.* 1978) occurred in a village in the USA where 37 of 119 people reporting ill with abdominal pain and fever were found to be infected with *Y. enterocolitica* type 0:8. The same serotype was also isolated from 1 of 60 people reporting diarrhoea without abdominal pain and fever. Twenty three of the culture positive patients, 9 of 40 culture negative and one symptomless child had a raised antibody level (>128). Sixteen patients underwent appendectomy. The common food source of the organism was identified as a chocolate milk drink. This was purchased from a small dairy which was the exclusive supplier of milk to the local schools. The chocolate milk was prepared by adding chocolate syrup to previously pasteurized milk in an open vat and mixing by hand with a perforated metal stirring rod. *Yersinia enterocolitica* type 0:8 was isolated from 1 of 4 unopened cartons of milk obtained from the school cafeteria.

Yersinia enterocolitica has been isolated from many sources including mussels (Spadaro & Infortuna 1968), drinking water (Lassen 1972), oysters (Toma 1973), poultry meat, pork and beef (Leistner *et al.* 1975) and also from milk (Schiemann & Toma 1978). Pigs, dogs, hares and chinchillas are known hosts of the organism.

The biochemical classification and differentiation of *Y. enterocolitica* and *Yersinia*-like organisms is still under debate but detailed schemes for characterization of these bacteria have been described by many workers (Wauters 1970; Darland *et al.* 1974; Sonnenwirth 1976; Mehlman *et al.* 1978).

Two biotyping schemes which sub-divide *Y. enterocolitica* into five biochemical groups have also been described (Niléhn 1969; Wauters 1970) and further sub-division of the species is possible using a serotyping scheme based on the somatic antigens (Wauters 1970) and by phage typing (Niléhn & Ericson 1969). There are 34 serotypes of which types 0:3, 0:8 and 0:9 have been most commonly associated with human illness. In a correlation study of serological and phage types it was found that most of the human serotypes were phage types II or III (Niléhn 1973).

This report is part of a larger study on the growth, survival and isolation of *Y. enterocolitica* in various foods.

Materials and Methods

Organisms

Two human isolates of *Y. enterocolitica* serotypes 0:3 and 0:9 were supplied by the Yersinia Reference Laboratory, Public Health Laboratory, Leicester.

Media

All dehydrated media were prepared and sterilized according to the manufacturers' instructions.

Brain Heart Infusion broth (BHI): Dehydrated Difco Bacto No. B37.

Brain Heart Infusion agar plates (BHI +dye): BHI with the addition of (% w/v) powdered agar 1·2; lactose 1; 1 % solution neutral red 0·5.

Buffered Brain Heart Infusion broth (BBHI): Sterile double strength BHI mixed with an equal volume of sterile 0·68 M McIlvaine's citric acid phosphate buffer solution of the required pH (Giegy Scientific Tables 1962).

GN broth (Hajna 1955) *(GN):* Dehydrated Difco Bacto No. 0486.

Phosphate buffered saline (PB): 0·067 M potassium dihydrogen phosphate/disodium hydrogen phosphate (pH 7·6) in 0·85 % (w/v) sodium chloride.

Quarter strength Ringer's solution (RS): Oxoid tablets No. BR 52.

Selenite broth (SB): (% w/v) sodium hydrogen selenite 0·4; peptone 0·5; mannitol 0·4; disodium hydrogen phosphate 0·75; sodium dihydrogen phosphate 0·53 in distilled water. Sterilized by filtration.

Selenite cystine broth (SC): Dehydrated Difco Bacto No. 0687.

Wauters' broth (Wauters 1973) *(WB):* Sterile 1 % Bacto tryptone No. B123 620 ml; sterile M/15 disodium hydrogen phosphate 160 ml; sterile 40 % magnesium chloride 208 ml; 0·2 % (w/v) malachite green 6·4 ml; 1000 μg ml $^{-1}$ carbenicillin 1·2 ml. The medium was stored at 4°C until required.

Foods

Hard-boiled egg, boiled fish, rice, potato, and roast chicken were prepared and cooked by normal domestic methods. After cooking, the bones and other non-edible parts were removed and where necessary the foods were chopped or finely homogenized. These foods were chosen primarily because of their previous implication in other types of food-borne outbreaks. Studies on other foods with a higher natural contaminating flora have still to be completed.

Growth studies

Preparation of inocula. Overnight BHI broth cultures incubated at 30°C were diluted with RS to give the required concentration for each experiment.

Surface colony counts. 10-fold dilutions were prepared in RS and counts made by a modified surface drop method (Thatcher & Clark 1968) on BHI plus dye plates.

Growth at different pH values: The two test strains were inoculated separately into 20 ml volumes of BBHI broth at pH 2·4, 3·6, 4·2, 4·4, 4·6, 4·8, 5·4, 6·6 and 7·8 to give approx. 10^3 colony forming units (cfu) ml $^{-1}$ and incubated at 22°C for 72 h. Surface colony counts were carried out after 0, 8, 14, 18, 22, 26, 30, 38, 48, 62 and 72 h and the plates incubated at 30°C for 48 h.

Growth in various media. Both test strains were inoculated separately into each of 3×100 ml volumes of BHI, WB, SB, SC, GN and PB to give approx. 100 cfu ml $^{-1}$. The broths were incubated at 4° and 22°C for 13 d and at 30°C for 72 h. Surface colony counts were carried out after 1, 2, 5, 9 and 13 days at 4°C and 0, 6, 24 and 48 h at 22° and 30°C; plates were incubated at 30°C for 48 h.

Growth in foods. 10 gram quantities of each homogenized food were distributed in sterile 1 lb screw-capped jars. 0·06 ml volumes of the two test strains were inoculated separately on to the surface of each of 20×10 g samples to give approx. 10^3 cfu g $^{-1}$. Five inoculated jars of food were incubated at 30°C, six at 4°C and seven at −20°C. One jar was retained and a surface colony count carried out immediately. The counts were carried out by homogenizing a 10 g sample with 90 ml RS to give a 1/10 dilution, from which further dilutions were prepared.

Surface colony counts were carried out on the foods incubated at 30°C after 5, 8, 14, 24 and 30 h, from those stored at 4°C after 1, 2, 3, 7, 10 and 21 d and at −20°C after 1, 2, 3, 21, 35, 49 and 63 d. All the plates were incubated at 30°C for 48 h.

Where it was suspected that counts would be <100 g $^{-1}$, and therefore beyond the lower limits of the counting method, 90 ml volumes of BHI broth were added to the 10 g portions of food instead of RS and these suspensions incubated at 30°C for 48 h as enrichment cultures after being used as 1/10 dilutions for the surface colony counts. Sub-cultures were made on BHI + dye plates and incubated at 30°C for 48 h. The presence or absence of *Y. enterocolitica* was recorded.

Results

Effect of pH

Figure 1 shows the growth curves of *Y. enterocolitica* serotype 0:3 at various pH values in BBHI; similar results were obtained with serotype 0:9. The lower the pH value the more prolonged was the lag period. At pH 3·6 and below there was a rapid decrease in the number of *Y. enterocolitica* and within 48 h the number of both strains had diminished to below the limits of the counting method. At pH 4·2 both strains sur-

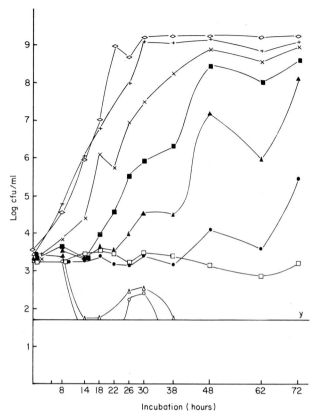

Fig. 1. Growth of *Y. enterocolitica* serotype 0:3 in BBHI at 22°C. ○, pH 2·4; △, pH 3·6; □, pH 4·2; ●, pH 4·4; ▲, pH 4·6; ■, pH 4·8, x, pH 5·4; +, pH 6·6; ◇, pH 7·8; y, lower limit of counting method.

vived for 72 h with little or no variation from the original count. pH values of 4·2–4·4 appeared to be the threshold of inhibition/growth and although at pH 4·4 the lag period was prolonged, growth did occur, particularly after 38 h, and finally reached a level of $>2·75 \times 10^5$ cfu ml^{-1} at 72 h. Growth of both strains occurred more readily at pH 4·8 and all values above this. For example at pH 6·6 and 7·8 levels of $>1 \times 10^9$ cfu ml^{-1} were attained within 30 h.

Growth in enrichment media

In BHI both strains of *Y. enterocolitica* grew to levels greater than 5×10^8 cfu ml^{-1} within 48 h at 22° and 30°C. At 4°C the count rose more steadily and finally reached $>5 \times 10^8$ cfu ml^{-1} after 9 d. BHI was used as a standard against which the growth in all the other media was compared. The growth patterns of the two test strains in GN were almost identical to those in BHI, the counts attained being 5×10^8 cfu ml^{-1} after 10–11 d at 4°C.

Although PBS does not provide any organic nutrients for growth both strains survived for 13 d at 4°C with no decrease in count. At 22°C, the number of organisms fell below the lower counting limit within 2–5 d.

The two selenite broths SB and SC demonstrated very different characteristics. Figure 2a shows the growth of both serotypes in SC at 4°, 22° and 30°C. By comparison with BHI there was inhibition of *Y. enterocolitica* as would be expected in a selective medium. At 4°C growth increased steadily to 2.75×10^6 and 5×10^5 cfu ml^{-1} respectively after 13 d. The experiment was terminated before the upper limit of bacterial growth was attained. Nevertheless, at 30°C counts of 2×10^6 and 2×10^7 and at 22°C 6.2×10^4 and 8×10^5 cfu ml^{-1} were obtained within 48–72 h.

SB failed to support the growth of serotype 0:3 at any temperature and this strain was undetectable by the methods used after the initial count at 0 h. Serotype 0:9 was also undetectable at 4°C after the initial counts but at 22° and 30°C numbers of this serotype rose to 1.2×10^5 and 1.5×10^8 cfu ml^{-1} respectively within 2 d.

In WB only one strain, serotype 0:9, appeared to survive at 4°C with evidence of some growth after 9 d; serotype 0:3 was not isolated after 1 d. Both strains grew well at 22° and 30°C to $>3 \times 10^7$ cfu ml^{-1} in 2 d (Fig. 2b).

Growth in foods

Growth patterns of *Y. enterocolitica* in hard boiled egg and boiled fish are representative of the results from all five foods (Fig. 3a & b). At 30°C the colony counts rose rapidly to $>1 \times 10^8$ cfu g^{-1} within 30 h. There was a steady increase in count at 4°C over a period of 10 d by which time the number of *Y. enterocolitica* had reached the same level as at 30°C. Both strains of *Y. enterocolitica* survived for at least 8–9 weeks after storage at -20°C, the number of viable cells remaining at the same level as the initial count.

Boiled potato was the only food tested which became discoloured by the growth of the *Y. enterocolitica*.

Discussion

The results obtained in this study for growth of *Y. enterocolitica* are comparable in many respects with those of Hanna *et al.* (1977a). These workers adjusted the pH of BHI with sodium hydroxide or hydrochloric

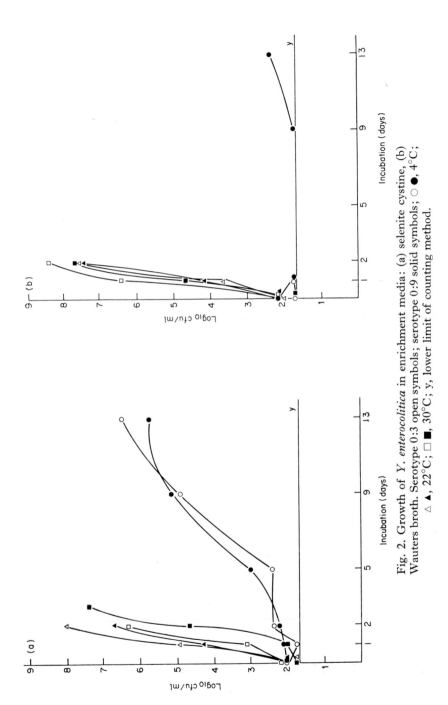

Fig. 2. Growth of *Y. enterocolitica* in enrichment media: (a) selenite cystine, (b) Wauters broth. Serotype 0:3 open symbols; serotype 0:9 solid symbols; ○ ●, 4°C; △ ▲, 22°C; □ ■, 30°C; y, lower limit of counting method.

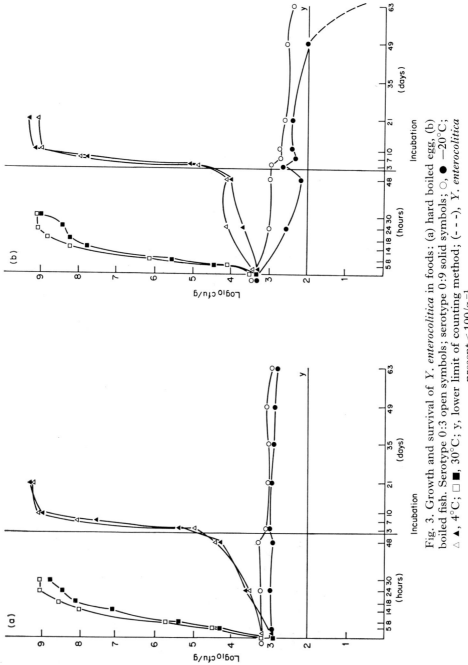

Fig. 3. Growth and survival of *Y. enterocolitica* in foods: (a) hard boiled egg, (b) boiled fish. Serotype 0:3 open symbols; serotype 0:9 solid symbols; $\bigcirc$, $-20°C$; $\triangle$ $\blacktriangle$, $4°C$; $\square$ $\blacksquare$, $30°C$; y, lower limit of counting method; (- -), *Y. enterocolitica* present $< 100/g^{-1}$.

acid prior to sterilization and only recorded the growth of the organisms over 24 h. The only difference between our results and those of Hanna and co-workers is that they found no increase in growth over 24 h at pH 5·0 whereas our results show a definite increase in numbers within that period. Our findings indicate that *Y. enterocolitica* can grow to high and therefore presumably hazardous levels in certain foods, even with a relatively low pH, when stored for prolonged periods.

Various methods have been recommended for the isolation of *Y. enterocolitica* but there appears to be little information in the literature relating to the performance of specific media. Toma & Diedrick (1975) used phosphate buffered saline pH 7·6 incubated at 4°C for the isolation of *Yersinia* from the caecal contents of pigs. Inoue & Kurose (1975) used both phosphate buffered saline and selenite broth with novobiocin when investigating cow's intestinal contents and beef meat. Of the 24 isolations obtained only one was made from selenite—novobiocin. Phosphate buffered saline was also used successfully in the investigation of the chocolate milk outbreak (Black *et al.* 1978). Schiemann & Toma (1978) compared four methods of isolating *Yersinia* from milk in which they included phosphate buffer and a modified Rappaport broth. The modified Rappaport broth was identical in formula to that of the Wauters broth used in this study. One millilitre of sample was inoculated into 10 ml of enrichment media as follows: (1) modified Rappaport broth incubated at room temperature for 5 d; (2) Butterfields phosphate buffer pH 7·2 incubated at 4°C for 14 d; (3) modified Rappaport broth inoculated with 1 ml of phosphate buffer culture (incubated at 4°C for 14 d) and incubated at 23°C for 5 d; and (4) modified Rappaport broth inoculated with 1 ml from a cooked meat culture (incubated at 23°C for 28 d and incubated at 23°C for 5 d. The four methods gave 6, 8, 24 and 3 isolations respectively.

Extended incubation at 4°C is widely used as an enrichment technique for *Y. enterocolitica* but not all media which support growth at 22° and 30°C are suitable for use in low temperature procedures. This appears to be true for Wauters broth as shown by the results presented in this paper. However when Schiemann & Toma (1978) used this medium at 22°C in combination with PB at 4°C (Method 3) isolations were substantially increased. Our results show that *Y. enterocolitica* survives in PB at 4°C but does not multiply without the addition of nutrients. The increased recovery of this organism by Schiemann & Toma's method 3 could have been due to *Y. enterocolitica* surviving better in the PB than the competing flora due to the presence of milk in the sample. Selection might have been further enhanced when some of the culture was transferred to modified Rappaport broth and incubated at 22°C for 5 d. The results obtained in this study with SB and SC show the former medium to be unsuitable as an enrichment medium for *Y. enterocolitica*. The main difference between the two media apart from the fact that SB contains mannitol and SC lactose is that SC also contains 0·001 % *l*-cystine.

There are few published reports of studies on the growth and survival of *Y. enterocolitica* in foods. Our results are similar to those obtained by Hanna *et al.* (1977b) for raw and cooked beef and pork. At 7°C the main increase in numbers occurred between 3 and 7 d storage. Hanna *et al.* (1977a) also examined the effect of freezing on the survival of *Y. enterocolitica* in cooked beef. They used two inoculum levels: 1×10^3 to 1×10^4 and 1×10^6 to 1×10^7 cfu g^{-1} and storage was at $-20°C$. One strain of *Y. enterocolitica* could not be detected after two weeks and two other strains could not be detected after four weeks. These results are quite different to those presented in this paper where very small numbers of *Y. enterocolitica* could still be demonstrated even after 9 weeks at $-20°C$.

In conclusion, the results presented in this paper indicate that *Y. enterocolitica* can grow to high and therefore presumably hazardous levels in foodstuffs stored at domestic refrigerator temperature (4°C). In a warm atmosphere (22°C) even those foodstuffs with the relatively low pH of 4·8, if kept for a prolonged period (48 h), could be a hazard.

Further investigations are needed to find the ideal technique for isolation of *Y. enterocolitica* and the enrichment broth/incubation temperature combination needs to be carefully selected.

We are grateful to Dr Diane Roberts for her helpful advice and criticism in the preparation of this paper.

References

ASAKAWA, Y., AKAHANE, S., KAGATA, N., NOGUCHI, M., SAKAZAKI, R. & TAMURA, K. 1973 Two community outbreaks of human infection with *Yersinia enterocolitica*. *Journal of Hygiene, Cambridge* **71**, 715–723.

BLACK, R. E., JACKSON, R. J., TSAI, T., MEDVESKY, M., SHAYEGANI, M., FEELEY, J. C., MACLEOD, K. I. E. & WAKELEE, A. M. 1978 Epidemic *Yersinia enterocolitica* infection due to contaminated chocolate milk. *New England Journal of Medicine* **298**, 76–79.

BOTTONE, E. J. 1977 *Yersinia enterocolitica* a panoramic view of a charismatic micro-organism. *CRC Critical Reviews in Microbiology* **5**, 211–241.

DARLAND, G., EWING, W. H. & DAVIS, B. R. 1974 The biochemical characteristics of *Yersinia enterocolitica* and *Yersinia pseudotuberculosis*. *US Department of Health, Education and Welfare* No. (CDC) 75–8294.

GEIGY SCIENTIFIC TABLES 1962 ed. Diem, K. pp. 314–315, Manchester: Geigy Pharmaceuticals.

HAJNA, A. A. 1955 A new enrichment broth medium for gram-negative organisms of the intestinal group. *Public Health Laboratory* **13**, 83–90.

HANNA, M. O., STEWART, J. C., CARPENTER, Z. L. & VANDERZANT, C. 1977*a* Effect of heating, freezing and pH on *Yersinia enterocolitica*-like organisms from meat. *Journal of Food Protection* **40**, 689–692.

HANNA, M. O., STEWART, J. C., ZINK, D. L., CARPENTER, Z. L. & VANDERZANT, C. 1977*b* Development of *Yersinia enterocolitica* on raw and cooked beef and pork at different temperatures. *Journal of Food Science* **42**, 1180–1184.

HIGHSMITH, A. K., FEELEY, J. C. & MORRIS, G. K. 1977 *Yersinia enterocolitica:* A review of the bacterium and recommended laboratory methodology. *Health Laboratory Science* **14**, 253–260.

HINDERAKER, S., LIAVAAG, I. & LASSEN, J. 1973 *Yersinia enterocolitica* infection. *Lancet II*, 322–323.

INOUE, M. & KUROSE, M. 1975 Isolation of *Yersinia enterocolitica* from cow's intestinal contents and beef meat. *Japanese Journal of Veterinary Science* **37**, 91–93.

KOHL, S., JACOBSON, J. A. & NAHMIAS, A. 1976 *Yersinia enterocolitica* infections in children. *Journal of Pediatrics* **89**, 77–79.

LASSEN, J. 1972 *Yersinia enterocolitica* in drinking-water. *Scandinavian Journal of Infectious Diseases* **4**, 125–127.

LEISTNER, L., HECHELMANN, H., KASHIWAZAKI, M. & ALBERTZ, R. 1975 Nachweis von *Yersinia enterocolitica* in faeces und fleisch von Schweinen, Rindern und Geflügel. *Fleischwirtschaft* **55**, 1599–1602.

MAIR, N. S. 1975 Yersiniosis (Infections due to *Yersinia pseudotuberculosis* and *Yersinia enterocolitica.*) In *Diseases Transmitted from Animals to Man.* eds Hubbert, W. T., McCulloch, W. F. & Schnurrenberger, P. R. Ch. 13, pp. 174–185, Illinois: Thomas.

MEHLMAN, I. J., AULISIO, C. C. G. & SANDERS, A. C. 1978 Problems in the recovery and identification of *Yersinia* from food. *Journal of the Association of Analytical Chemists* **61**, 761–771.

MORRIS, G. K. & FEELEY, J. C. 1976 *Yersinia enterocolitica*: a review of its role in food hygiene. *Bulletin of the World Health Organisation* **54**, 79–85.

NILÉHN, B. 1969 Studies on *Yersinia enterocolitica* with special reference to bacterial diagnosis and occurrence in human enteric disease. *Acta Pathologica et Microbiologica Scandinavica* Supplement 206.

NILÉHN, B. 1973 Host range, temperature characteristics and serologic relationships among *Yersinia* phages. In *Contributions to Microbiology and Immunology,* Yersinia, Pasteurella *and* Francisella ed. Winblad, S. Vol. 2, pp. 59–67, Basel: Karger.

NILÉHN, B. & ERICSON, C. 1969 Studies on *Yersinia enterocolitica*. Bacteriophages liberated from chloroform treated cultures. *Acta Pathologica et Microbiologica Scandinavica* **75**, 117–187.

OLŠOVSKÝ, Z., OLŠÁKOVÁ, V., CHOBOT, S. & SVIRIDOV, V. 1975 Mass occurence of *Yersinia enterocolitica* in two establishments of collective care of children. *Journal of Hygiene, Epidemiology, Microbiology and Immunology* **19**, 22–29.

RABSON, A. R. & KOORNHOF, H. J. 1972 *Yersinia enterocolitica* infections in South Africa. *South African Medical Journal* **46**, 798–803.

RAKOVSKÝ, J., PAUČKOVÁ, V. & ALDOVÁ, E. 1973 Human *Yersinia enterocolitica* infections in Czechoslovakia. In *Contributions to Microbiology and Immunology,* Yersinia, Pasteurella *and* Francisella ed. Winblad, S. Vol. 2, pp. 93–98, Basel: Karger.

SCHIEMANN, D. A. & TOMA, S. 1978 Isolation of *Yersinia enterocolitica* from raw milk. *Applied and Environmental Microbiology* **35**, 54–58.

SONNENWIRTH, A. C. 1976 Isolation and characterization of *Yersinia enterocolitica*. *Mount Sinai Journal of Medicine* **43**, 736–745.

SPADARO, M. & INFORTUNA, M. 1968 Isolamento di *Yersinia enterocolitica* in *Mitilus galloprovincialis* lamk. *Bollettino della Società Italiana di Biologia Sperimentale* **44**, 1896–1897.

SZITA, J. & SVIDRÓ, A. 1976 A five year study of human *Yersinia enterocolitica* infections in Hungary. *Acta Microbiologica Academiae Scientiarum Hungaricae* **23**, 191–203.

THATCHER, F. S. & CLARK, D. S. 1968 *Microorganisms in Foods. Their Significance and Methods of Enumeration.* Toronto: University of Toronto Press.

TOMA, S. 1973 Survey on the incidence of *Yersinia enterocolitica* in the province of Ontario. *Canadian Journal of Public Health* **64**, 477–487.

TOMA, S. & DIEDRICK, V. R. 1975 Isolation of *Yersinia enterocolitica* from swine. *Journal of Clinical Microbiology* **2**, 478–481.

VANDEPITTE, J., WAUTERS, G. & ISEBAERT, A. 1973 Epidemiology of *Yersinia enterocolitica* infections in Belgium. In *Contributions to Microbiology and Immunology*, Yersinia, Pasteurella *and* Francisella ed. Winblad, S. Vol. 2, pp. 111–119, Basel: Karger.

WAUTERS, G. 1970 *Contribution a l'étude de* Yersinia enterocolitica. *Thesis*, University of Louvain.

WAUTERS, G. 1973 Improved methods for the isolation and recognition of *Yersinia enterocolitica*. In *Contribution to Microbiology and Immunology*, Yersinia, Pasteurella *and* Francisella ed. Winblad, S. Vol. 2, pp. 68–70, Basel: Karger.

ZEN-YOJI, H. & MARUYAMA, T. 1972 The first successful isolations and identification of *Yersinia enterocolitica* from human cases in Japan. *Japanese Journal of Microbiology* **16**, 493–500.

ZEN-YOJI, H., MARUYAMA, T., SAKAI, S., KIMURA, S., MIZUNO, T. & MAMOSE, T. 1973 An outbreak of enteritis due to *Yersinia enterocolitica* occurring at a Junior High School. *Japanese Journal of Microbiology* **17**, 220–222.

Influence of Sodium Chloride, pH and Temperature on the Inhibitory Activity of Sodium Nitrite on *Listeria monocytogenes*

M. Shahamat, A. Seaman and M. Woodbine

Microbiological Unit, Department of Applied Biochemistry and Nutrition, University of Nottingham School of Agriculture, Sutton Bonington, Loughborough, Leicestershire, UK

The inhibitory effect of sodium nitrite on food poisoning bacteria has been studied by many workers using different laboratory media as well as food products under different conditions (e.g. Sair 1971; Baird-Parker & Baillie 1973; Ashworth *et al.* 1974; Roberts & Smart 1974; Bean & Roberts 1974; Roberts 1975; Crowther *et al.* 1976; Jarvis *et al.* 1976; Labots 1976; Mirna & Coretti 1976; Roberts & Derrick 1978). In practice, nitrite is said to have relatively little inhibitory effect on bacteria unless accompanied by other factors such as sodium chloride and low pH. However, the effect of sodium nitrite, or the interaction between sodium chloride, sodium nitrite, pH and temperature, on *Listeria monocytogenes* have not been reported.

Listeria monocytogenes can survive in, and be disseminated by, different kinds of food such as milk, poultry, eggs and meat. It can survive most salting procedures used in food processing and can also survive at low temperature in meat (Khan *et al.* 1973; Khan *et al.* 1975), which makes it a potential public health hazard. *Listeria monocytogenes* is listed among the psychrotrophic food spoilage micro-organisms (Mossel 1971). The studies undertaken here, therefore, were to investigate the effect of various combinations of sodium nitrite and sodium chloride at different temperatures and at different pH levels on the growth of *L. monocytogenes*.

Materials and Methods

Organism and media

Listeria monocytogenes 1/2a (No. 18) haemolytic strain (B.S. Ralovich, Institute of Public Health and Epidemiology, University Medical School,

Pecs, Hungary) was used. The organism was maintained on trypticase soy agar (B.B.L.), stored in a refrigerator and subcultured monthly. The basal medium used throughout was typticase soy broth (TSB) (trypticase-peptone 1·7%, phytone-peptone 0·3%, sodium chloride 0·5%, dipotassium phosphate 0·25% and dextrose 0·25%). The desired amount of sodium chloride (AR; calculated as % w/v) was dissolved in the medium which was then dispensed as 200 ml quantities into 500 ml Winchester bottles. When sodium nitrite (AR) was required, solutions of different concentrations were prepared in distilled water which was then added to each bottle in 10 ml quantities to give the desired final concentration. The pH of each medium was then adjusted using concentrated HCl (AR) to the approximate pH end-point followed by diluted HCl to 0·1 u above the desired value (using a Pye Unicam PW 418 pH meter fitted with a glass electrode). The media were autoclaved at 121°C for 15 min. The pH of each medium was checked before inoculation and at the end of each experiment.

Inoculation procedure

For inoculation the organism was first grown from stock on the surface of a TSA plate. After 48 h incubation at 37°C this was seeded into a 1 oz universal screw cap bottle containing 10 ml TSB. An 18 h culture was used for inoculation into Winchester bottles. Checks were made by absorbance and viable count that inocula were approximately equal.

Counting techniques

Viable plate counts were made by inoculating and evenly distributing 0·5 ml of diluted bacterial suspension using overlay methods. The TSA plates were dried by 24 h incubation at 37°C with the lids closed and 2 h incubation at 37°C with the lids and base separated. Bacterial suspensions were diluted in freshly made 0·85% saline solution. Colonies were counted after 48 h incubation at 37°C.

Growth measurements

Absorbance was measured using a Unicam Spectrophotometer SP 600 series 2 (Pye Unicam Ltd., Cambridge). Measurements were made at 650 nm on 3·5 ml volumes of culture with the basal medium as blank.

Growth conditions

Incubation temperatures used were 4, 22 and 37°C. All the experiments were repeated twice except those at extreme conditions which were repeated three or four times.

Analytical methods

To evaluate the fate of nitrite in media during heating and during long incubation at low temperatures (4°C), the amount of nitrite was determined at five different times: immediately after dissolving in the media; after seven days incubation at 4°C; after 30 days incubation at 4°C; after 60 days incubation at 4°C, which was the end of the experiment. The assay was made following the methods of modified Griess reagent (A.O.A.C. 1975).

Results

The cultures of *L. monocytogenes* were grown at 4,22 and 37°C in basal medium (TSB) and in basal medium plus different concentrations of salts at various pH values. Growth was measured over a 15 day period at 22 and 37°C and over a 60 day period at 4°C. The influence of temperature on the inhibitory activity of sodium nitrite at pH 7·4 is shown in Fig. 1. Each graph in this series shows closed symbols for the maximum absorbance of the organism in each concentration of nitrite, expressed as a percentage of the maximum absorbance of the organism in the basal

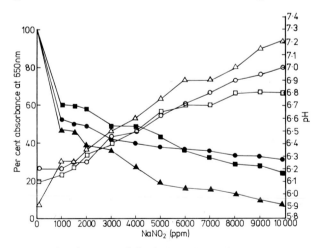

Fig. 1. Maximum absorbance and final pH values of *L. monocytogenes* grown at pH 7·4 at varying nitrite concentrations. □ ■, 37°C; ○ ●, 22°C; △ ▲, 4°C; □ △ ○, pH change; ■ ▲ ●, growth.

medium without nitrite. The open symbols show the final reduction in the pH of the media and each point indicates the final reduction in pH in each medium with a particular concentration of nitrite. The organism was able to grow in all the concentrations of nitrite tested at the three incubation temperatures, but the amount of growth was reduced as the amount of nitrite increased. The relation between growth and reduction of pH seems to be direct. The influence of temperature on the in-hibitory effects of nitrite at lower pH (6·5) is shown in Fig. 2. The

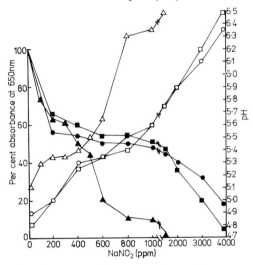

Fig. 2. Maximum absorbance and final pH values of *L. monocytogenes* grown at pH 6·5 at varying nitrite concentrations. □ ■, 37°C; ○ ●, 22°C; △ ▲, 4°C; □ ○ △, pH; ■ ● ▲, growth.

inhibitory effect is stronger at this pH and a temperature of 4°C had great influence on the action of nitrite, which was inhibitory at a con-centration of 1500 p.p.m. The influence of temperature on the in-hibitory effects of nitrite at pH 5·5 is shown in Fig. 3. At this low pH the minimum inhibitory concentration of nitrite was reduced at all three incubation temperatures. Nevertheless, the inhibitory action was greatest again at the lowest incubation temperature, when 600 p.p.m. was suffi-cient to suppress growth. The influence of temperature on the inhibitory actions of nitrite at the lowest pH (5), at which growth could occur in the basal medium is illustrated in Fig. 4. The effect of nitrite at 4°C is not shown because this organism cannot grow at this temperature when the pH is reduced to 5·0. The bacteriostatic action of sodium nitrite on *L. monocytogenes* was strongest at this pH at both temperatures. These data are summarized in Table 1.

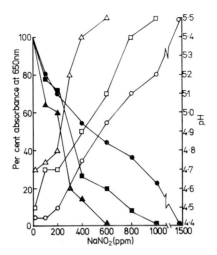

Fig. 3. Maximum absorbance and final pH values of *L. monocytogenes* grown at pH 5·5 at varying nitrite concentrations. □ ■, 37°C; ○ ●, 22°C; △ ▲, 4°C; □ ○ △, pH; ■ ● ▲, growth.

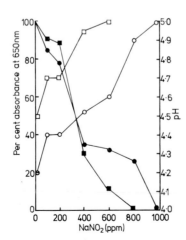

Fig. 4. Maximum absorbance and final pH values of *L. monocytogenes* grown at pH 5·0 at varying nitrite concentrations. □ ■, 37°C; ○ ●, 22°C; □ ○, pH; ■ ●, growth.

TABLE 1

Highest concentration of sodium nitrite (p.p.m.) for L. monocytogenes *No. 18 at which growth occurred under the influence of
temperature and pH*

Temperature	pH			
	7·4	6·5	5·5	5·0
37°C	25 000	3000	800	400
22°C	30 000	4000	1000	800
4°C	10 000	1000	400	...

To investigate possible synergistic effects, combinations of sodium
nitrite (ranging from 0 to 10 000 p.p.m.), sodium chloride (0·5, 3, 5·5
and 8%), pH (5·0, 5·5, 6·5 and 7·4) and different incubation temperatures
(4, 22 and 37°C) were used. The aim of these experiments was to estab-
lish the growth and inhibition zone for the organism under extreme
environmental conditions. It is clear from the results in Table 2 that
sodium chloride had a great influence on the bacteriostatic action of
sodium nitrite. The depletion of nitrite during heat processing and
storage is thought to be relevant when it is introduced into meat products.
In laboratory media, however, this reduction depends on the nature of
the medium, pH, and storage temperature. Perigo *et al.* (1967) found

TABLE 2

Minimum inhibitory concentrations of sodium nitrite (p.p.m.) for
L. monocytogenes *No. 18 under the influence of salt, pH and
temperature*

Temperature	pH	NaCl (%)		
		3	5·5	8
	7·4	10 000	10 000	5000
	6·5	5000	3000	1500
37°C				
	5·5	1000	500	<50
	5·0	500	<50	...
	7·4	10 000	10 000	7000
	6·5	5000	3000	1500
22°C				
	5·5	1000	1000	<50
	5·0	500	<50	...
	7·4	3000	2000	...
	6·5	600	300	...
4°C				
	5·5	100	<50	...

that a significant amount of nitrite added to raw meat products was not detectable after heat processing. This disappearance of nitrite is much greater when it is heated in meat than when heated to the same degree in water. Bayne & Michener (1975) found that the nitrite concentration in broth remained approximately constant during their experiments. In the experiments described here nitrite concentrations in all media stored at 4°C remained approximately constant during the course of the experiments.

Discussion

Since Tarr (1942) concluded that the inhibitory action of nitrite for a variety of bacteria in fish muscle or bacteriological medium depends on the pH value, numerous papers have been published comparing the influence of pH and other factors such as sodium chloride and temperature on the action of nitrite on different organisms—particularly those which are involved with food production.

Castellani & Niven (1955) reported that the pH of the medium had a great influence on the bacteriostatic action of nitrite in which the minimum inhibitory concentration of nitrite necessary to suppress the growth of *Staph. aureus* was 4000 p.p.m. at pH 6·9 but was reduced to only 80 p.p.m. when the pH was 5·05. They concluded that as the pH of the medium is reduced by one unit the bacteriostatic action of nitrite is increased approximately 10-fold. Buchanan & Solberg (1972) found that when *Staph. aureus* was exposed to sodium nitrite at pH 7·3 the growth did not cease up to 2000 p.p.m., which is 10 times the level permitted in cured meat products. When the pH was reduced to 6·3, however, the effect of nitrite was enhanced and *Staph. aureus* was destroyed by ≥500 p.p.m. Bayne & Michener (1975) tested nitrite concentrations from 0 to 8000 p.p.m. in a tryptone broth medium against *S. enteritidis* and *S. typhimurium* at pH values from 4·6 to 7·5, and found the inhibitory effect of nitrite was strongly pH dependent and, with the concentration permitted in meat, nitrite was ineffective above pH 6·0. Some other authors (Duncan & Foster 1968; Grever 1973; Roberts & Garcia 1973) also demonstrated the relation between the inhibitory action of nitrite and the pH of the medium.

In the present experiments pH had a similarly great influence on the inhibitory action of nitrite against *L. monocytogenes* (Fig. 5); even concentrations as high as 25 000 p.p.m. could not prevent growth at pH 7·4. In fact *L. monocytogenes* reacted to nitrite under these conditions similarly to *Lactobacillus arabinosus* in Castellani & Niven (1955) medium where the minimum inhibitory concentration of nitrite was 25 000 p.p.m., but with a pH value of 6·6. The inhibitory effect was greatly enhanced by

M. SHAHAMAT *ET AL.*

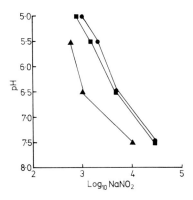

Fig. 5. M.I.C. of $NaNO_2$ ($\log_{10}$ p.p.m.) as a function of pH and temperature for *L. monocytogenes*. ■, 37°C; ●, 22°C; ▲, 4°C.

low pH and by temperature so that when the organism was exposed to nitrite at the same pH but at a lower temperature, the bacteriostatic action was enhanced (Table 1). Some authors (Segner *et al.* 1966; Baird-Parker & Freame 1967; Ohye & Christian 1967; Ohye *et al.* 1967) have also suggested that lower concentrations of salts inhibit the growth of micro-organisms at lower temperatures. Spencer (1967) pointed out that the concentration of curing salts necessary to prevent growth and toxin production by *Clostridium botulinum* is related to the number of cells present and also to the storage temperature. Although the combined effects of nitrite, pH and temperature on *L. monocytogenes* is clear, and particularly the relation between pH and the minimum inhibitory concentrations of nitrite is linear, nevertheless, none of these minimum inhibitory concentrations at any of the pH values or temperatures studied are acceptable in food under the regulations which were introduced in 1974 (Jarvis & Burke 1976).

It was necessary, therefore, to take into account a fourth factor, which interacted in the systems: the addition of salt. Interaction between nitrite, salt and pH has been the concern of many workers. Roberts & Ingram (1973) emphasized that the inhibition of growth of *Cl. botulinum* types A, B, E and F is the result of interactions between pH, salt and nitrite. Therefore, it is misleading to consider modifying one of these components without taking the other two into account.

In the present experiments the interaction between these factors seems to be linear, consequently, the minimum inhibitory concentrations of salts were reduced to the lowest levels at the lowest pH values (5). The experiments strongly support the idea that the three factors: salt, nitrite and pH are equally important and nitrite cannot be considered by itself

(Ingram 1976). It should be emphasized that the synergistic aspect of the three factors (salt, nitrite and pH) has to be set against the absence of even a bacteriostatic effect of nitrite alone. The experiments confirmed the statement by Ingram (1976): "Many pH, salt, nitrite combinations which readily permit growth under warm conditions, are wholly inhibitory at lower temperatures". The results from these experiments also support the conclusions of Baird-Parker & Baillie (1973) that less nitrite and salt are required to inhibit the growth of *Cl. botulinum* types A, B, E and F at lower than at higher storage temperatures. These results clearly show the antimicrobial activity of nitrite against *L. monocytogenes* at a level within current food regulations can be achieved only in the presence of at least 3% NaCl and a pH value at or below 5·5 at cold storage temperatures. Ingram (1973) postulated that the optimum pH for the antimicrobial activity of nitrite is about 5·5. It thus seems that this pH is acceptable for controlling *L. monocytogenes*, which is amongst the psychrotrophic food spoilage flora (Mossel 1971), at low storage temperatures.

We thank Mr David Fowler for his technical assistance.

References

A.O.A.C. 1975 *Official Methods of Analysis of the Association of Official Analytical Chemists* ed. W. Horwitz. 12th edn, pp. 228, 422.

ASHWORTH, J., DIDCOCK, A., HARGREAVES, L. L., JARVIS, B. & WALTERS, C. L. 1974 Chemical and microbiological comparisons of inhibitors derived thermally from nitrite with an iron thionitrosyl (Roussin Black Salt). *Journal of General Microbiology* **84**, 403–408.

BAIRD-PARKER, A. C. & BAILLIE, M. A. H. 1973 The inhibition of *Clostridium botulinum* by nitrite and sodium chloride. In *Nitrite in Meat Products*. Proceedings of the International Symposium on Nitrite and Meat Products. Zeist, The Netherlands eds Krol, B. & Tinbergen, B. J. pp. 77–90, Wageningen: Pudoc.

BAIRD-PARKER, A. C. & FREAME, B. 1967 Combined effect of water activity, pH and temperature on the growth of *Clostridium botulinum* from spore and vegetative cell inocula. *Journal of Applied Bacteriology* **30**, 420–429.

BAYNE, H. G. & MICHENER, D. 1975 Growth of *Staphylococcus* and *Salmonella* on frankfurters with and without sodium nitrite. *Applied Microbiology* **30**, 844–849.

BEAN, P. G. & ROBERTS, T. A. 1974 The effect of pH, sodium chloride and sodium nitrite on heat resistance of *Staph. aureus* and growth of damaged cells in laboratory media. *Proceedings of the IV International Congress of Food Science and Technology*. Volume 111, pp. 93–102, Valencia, Spain.

BUCHANAN, R. L. & SOLBERG, M. 1972 Interaction of sodium nitrite, oxygen and pH on growth of *Staph. aureus*. *Journal of Food Science* **37**, 81–85.

CASTELLANI, A. G. & NIVEN, C. F. 1955 Factors affecting the bacteriostatic action of sodium nitrite. *Applied Microbiology* **3**, 154–159.

CROWTHER, J. S., HOLBROOK, R., BAIRD-PARKER, A. C. & AUSTIN, B. L. 1976 Role of nitrite and ascorbate in the microbiological safety of vacuum-packed sliced bacon. In *Nitrite in Meat Products*. Proceedings of the II International Symposium on Nitrite in Meat Products. Zeist, The Netherlands eds. Tinbergen, B. J. & Krol, B. pp. 13–20, Wageningen: Pudoc.

DUNCAN, C. L. & FOSTER, E. M. 1968 Role of curing agents in the preservation of shelf stable canned meat products. *Applied Microbiology*. **16**, 401–405.

GREVER, A. B. G. 1973 Minimum nitrite concentrations for inhibition of *Clostridia* in cooked meat products. In *Nitrite in Meat Products*. Proceedings of the II International Symposium on Nitrite in Meat Products. Zeist, The Netherlands eds Krol, B. & Tinbergen, B. J. pp. 103–109, Wageningen: Pudoc.

INGRAM, M. 1973 The microbiological effects of nitrite. In *Nitrite in Meat Products*. Proceedings of the II International Symposium on Nitrite in Meat Products. Zeist, The Netherlands eds Krol, B. & Tinbergen, B. J. pp. 63–76, Wageningen: Pudoc.

INGRAM, M. 1976 The microbiological role of nitrite in meat products. In *Microbiology in Agriculture, Fisheries and Food* eds. Skinner, F. A. & Carr, J. G. pp. 1–16. The Society for Applied Bacteriology, Symposium Series No. 4, London: Academic Press.

JARVIS, B. & BURKE, CAROLE S. 1976 Practical and legislative aspects of the chemical preservation of food. In *Inhibition and Inactivation of Vegetative Microbes* eds Skinner, F. A. & Hugo, W. B. pp. 345–367, The Society for Applied Bacteriology, Symposium Series No. 5, London: Academic Press.

JARVIS, B., RHODES, A. C., KING, S. E. & PATEL, M. 1976 Sensitization of heat-damaged spores of *Clostridium botulinum* type B to sodium chloride and sodium nitrite. *Journal of Food Technology* **11**, 41–50.

KHAN, M. A., PALMAS, C. V., SEAMAN, A. & WOODBINE, M. 1973 Survival versus growth of a facultative psychrotroph in meat and products of meat. *Zentralblatt für Bakteriologie, Parasitenkunde, Infectionskrankheiten und Hygiene Abteilung* II, **157**, 277–282.

KHAN, M. A., NEWTON, I. A., SEAMAN, A. & WOODBINE, M. 1975 Survival of *Listeria monocytogenes* inside and outside its host. In *Problems of Listeriosis*. Proceedings of the VI International Symposium in Problems of Listeriosis ed. Woodbine, M. pp. 75–83, Leicester: Leicester University Press.

LABOTS, H. 1976 Effect of nitrite on the development of *Staph. aureus* in fermented sausages. In *Nitrite in Meat Products*, Proceedings of the II International Symposium on Nitrite in Meat Products. Zeist, The Netherlands. eds Tinbergen, B. J. & Krol, B. pp. 21–27, Wageningen: Pudoc.

MIRNA, A. & CORETTI, K. 1976 Inhibitory effect of nitrite reaction products and of degradation products of food additives. In *Nitrite in Meat Products*. Proceedings of the II International Symposium on Nitrite in Meat Products. Zeist, The Netherlands. eds Tinbergen, B. J. & Krol, B. pp. 39–45, Wageningen: Pudoc.

MOSSEL, D. A. A. 1971 Ecological essentials of antimicrobial food preservation. Microbes and biological productivity. *Symposium of the Society for General Microbiology* **21**, 177–195.

OHYE, D. F. & CHRISTIAN, J. H. B. 1967 Combined effects of temperature, pH and water activity on growth and toxin production by *Cl. botulinum* types A,

B and E. In *Botulism 1966*. Proceedings of the V International Symposium on Food Microbiology, Moscow. eds Ingram, M. & Roberts, T. A. pp. 217–221.

OHYE, D. F., CHRISTIAN, J. H. B. & SCOTT, W. J. 1967 Influence of temperature on the water relations of growth of *Cl. botulinum* type E. In *Botulism 1966*. Proceedings of the V International Symposium on Food Microbiology, Moscow eds Ingram, M. & Roberts, T. A. pp. 136–142.

PERIGO, J. A., WHITING, E. & BASHFORD, T. E. 1967 Observations on the inhibition of vegetative cells of *Clostridium sporogenes* by nitrite which has been autoclaved in a laboratory medium, discussed in the context of sub-lethally processed cured meats. *Journal of Food Technology* 2, 377–397.

ROBERTS, T. A. 1975 The microbiological role of nitrite and nitrate. *Journal of the Science of Food and Agriculture* 26, 1755–1760.

ROBERTS, T. A. & DERRICK, C. M. 1978 The effect of curing salts on the growth of *Clostridium perfringens* (*welchii*) in a laboratory medium. *Journal of Food Technology* 13, 349–353.

ROBERTS, T. A. & GARCIA, C. A. 1973 A note on the resistance of *Bacillus* spp., faecal *Streptococci* and *Salmonella typhimurium* to an inhibitor of *Clostridium* spp. formed by heating sodium nitrite. *Journal of Food Technology* 8, 463–466.

ROBERTS, T. A. & INGRAM, M. 1973 Inhibition of growth of *Clostridium botulinum* of different pH values by sodium chloride and sodium nitrite. *Journal of Food Technology* 8, 467–475.

ROBERTS, T. A. & SMART, J. A. 1974 Inhibition of spores of *Clostridium* spp. by sodium nitrite. *Journal of Applied Bacteriology* 37, 261–264.

SAIR, L. 1971 Cure accelerators. *Proceedings of the XVII European Meeting of Meat Research Workers*. Bristol, UK pp. 201–211.

SEGNER, W. P., SCHMIDT, C. F. & BOLTZ, J. K. 1966 Effect of sodium chloride and pH on the outgrowth of spores of type E *Clostridium botulinum* at optimal and suboptimal temperatures. *Applied Microbiology* 14, 49–54.

SPENCER, R. 1967 Factors in curing meat and fish products affecting spore germination, growth and toxin production. In *Botulism 1966*, Proceedings of the V International Symposium on Food Microbiology, Moscow eds Ingram, M. & Roberts, T. A. pp. 123–135.

TARR, H. L. A. 1942. The action of nitrites on bacteria, further experiments. *Journal of the Fisheries Research Board of Canada* 6, 74.

Subject Index

Absidia spp., 67
Absidia corymbifera, 58, 59, 65
Acid rain, 22, 26
Actinomadura, 66
Actinomycetes, 55
  factors affecting growth, 56–59
  pH optima, 63
  temperature optima, 65
Aeration, 56–57
Agrobacterium sp., 28
Algae, blue-green, 28–29
Alginate bead technique, 182–185
Alkaline environments, 27, 28–29, 30
Alkaline springs, 27, 28
Alkalophiles, 27–36
  definition of, 28–29
  enrichment and isolation, 29–34
  halophilic, 33–34
  heterotrophic, 30
  pH optima, 34–36
  photosynthetic, 29, 30–32
Alkalotolerant organisms, 28, 34–35
Allescheria terrestris, 63
Alternaria sp., 64
Amino acids
  effect on deplasmolysis, 143–146
  and osmoregulation, 141–142
Anabaena sp.
  cell differentiation, 13
  colony formation, 10
Anabaenopsis sp., 29
Anaerobic bacteria, on solids, 197, 199–200
Anaerobic glove box system, 32
Ancalomicrobium sp., 2, 7
  enrichment and isolation, 3–4
  prostheca length, 5
Antibiotics in measurement of soil respiration, 25–26
Antimicrobial agents, 101
  in food, 122–124
Aspergillus spp., 56, 58, 59, 63–65, 67

Aspergillus cristatus, 134
Aspergillus fumigatus, 66
Aspergillus glaucus, 133–134
Aspergillus niger, 135
Aspergillus oryzae, 66
Asticcacaulis sp., 2
  enrichment and isolation, 4
Asticcacaulis biprosthecum, physiology of prosthecae, 8
ATP-ase inhibitors, 80–82

Bacillus spp., 28–29
  sporulation of, 161–162, 164, 166
Bacillus calfactor, 59, 66
Bacillus licheniformis, 66
Bacillus megaterium spores, in talc, 198, 203–204
Bacillus stearothermophilus, 59, 66, 189, 192
Bacillus stearothermophilus spores, 159
  in alginate bead technique, 182–185
  in capillary bulb technique, 180–182
  in capillary tube technique, 177–178
  heat resistance of, 174–175
  in laboratory scale UHT processing, 179–180
  in reconstituted food particle technique, 185–187
Bacillus subtilis, 83–84
Bacteria, *see also* individual organisms
  with acellular extensions, 16–18
  acidophilic, 39–48
  alkalophilic, 27–36
  anaerobic, on solids, 197, 199–200
  chemolithotropic, 40–42
  halophilic, 33–34, 56
  heterotrophic, 30, 42
  pH optima for growth, 63
  photosynthetic, 29–32
  prosthecate, 1–19
  respiration of, in soil, 24–26
  thermophilic, 39–48, 58

Food—Cont.
fermentation of, 66, 113, 118
preservatives, 122–124
Food spoilage, 104, 106–107, 112–113, 118
and water activity, 149
by xerophilic fungi, 129–130
Freshwater environments, microbial growth in,
see Prosthecate bacteria
Freshwater lakes, 10
Fungi
acid tolerant, 44
as health hazards, 67
pH optima, 63
in soil respiration, 24–26
in spontaneously heated materials, 55–57, 59
thermophilic, 58
xerophilic, 103, 109

Gamma-irradiation of solids, 197–198
Glucose amended soils, 24–26
Glycerol
and phenol concentration coefficient, 149, 150–155
and plasmolysis, 155
H$^+$ concentration, 71–72, see also pH
Halobacterium sp., 33
Halophilic alkalophiles, 33–34
Hansenula spp., 106–107, 115, 116
Hay, 53, 59–63
colonization and spontaneous heating of, 64–66
Heat sterilization, 173–187
Heat stress,
effect on growth of Yersinia enterocolitica, 209–212
High temperatures, inactivation of bacterial spores at, 173–187
Human gastrointestinal infection with Yersinia enterocolitica, 205, 215–216
Humectants, 124
Humicola lanuginosa, 59, 63, 65
Hydrostatic systems, 99–100
Hyperbaric oxygen, effect on Pseudomonas aeruginosa, 100–101
Hyperbaric systems, 100
Hyphomicrobium sp., 2, 5
dispersal, 13–16

enrichment and isolation, 2–3
growth of, 9–12
prostheca length, 5
Hysterium pulicare, 21

Infra-red gas analysis, 22–23
Ionizing radiation, resistance of microorganisms to, 195–204
materials and methods, 195–198
Iron, as energy source, 40–43, 45–47

Kakabekia barghoorniana, 28

Lactobacillus arabinosus, reaction to nitrite, 233–234
Lactose, effect on growth of Yersinia enterocolitica, 208, 210–212
Leach dump environments, 40, 44
Leptospirillum ferrooxidans, 41
Light microscopy, of prosthecate bacteria, 4
Listeria monocytogenes, 227–235
culture and growth, 227–229
inhibition by interacting factors, 229–235
Lycopodium, γ-irradiation of, 196–202

Maize starch, γ-irradiation of, 195–202
Malbranchea pulchella, 59, 63
Metal ions
culture on, 44–46
as energy sources, 40–44, 47
tolerance of organisms to, 43–44
Metal leaching environments, 40, 43–44
Micropolyspora faeni, 64–66
Milk, Yersinia enterocolitica in, 205–212
Monascus bisporus, 56, 130
Most probable number technique, 46–48
Mucor miehei, 63
Mucor pusillus, 66, 67
Mutants, with energy-linked metabolic defects, 92–96
isolation procedure, 94–96
in low pH resistance, 93–94
nomenclature, 92–93

Nitrite
bacteriostatic action, 233–235
effect on growth of Listeria monocytogenes, 229–235